Gathering Rare Ores

JONATHAN E. HELMREICH

Gathering Rare Ores

The Diplomacy of Uranium Acquisition, 1943-1954

Princeton University Press

Library of Congress Cataloging in Publication Data
will be found on the last printed page of this book
ISBN 0-691-04738-3

This book has been composed in Linotron Primer

Clothbound editions of Princeton University Press books
are printed on acid-free paper, and binding materials are
chosen for strength and durability

Printed in the United States of America
by Princeton University Press,
Princeton, New Jersey

To Nancy

CONTENTS

Contents

LIST OF ABBREVIATIONS

AEC	Atomic Energy Commission
CDA	Combined Development Agency, the successor in January 1948 to the
CDT	Combined Development Trust
CIA	Central Intelligence Agency (U.S.)
CPC	Combined Policy Committee
ERP	European Recovery Program
GOI	Government of India
MAUD Committee	Uranium subcommittee of Committee for the Scientific Survey of Air Warfare (U.K.)
MDAP	Mutual Defense Assistance Program
MED	Manhattan Engineering District
MLC	Military Liaison Committee (between the U.S. Joint Congressional Committee on Atomic Energy Affairs and the AEC)
MTU	Metric Tons of Uranium
NATO	North Atlantic Treaty Organization
NDRC	National Defense Research Committee (U.S.)
NSC	National Security Council (U.S.)
OCB	Operations Coordinating Board (U.S.)
S-1	Uranium (U.S.)
S-1 Committee	Office of Scientific Research and Development Section on Uranium (U.S.)
SGB	Société Générale de Belgique
UM	Union Minière du Haut Katanga
UMDC	United Mines Development Corporation
UN	United Nations

A premise of paucity undergirded the campaign of the United States, abetted by Great Britain and Canada, to gain a monopoly of uranium and thorium supplies during and immediately following the second world war. The falsity of that premise did not become clear until the early or middle 1950s; that discovery, the result of the detection of numerous sources of uranium and means to utilize low-grade ores, brought the major aspects of the diplomatic effort to an end.

The legacy of the venture was not negligible. It led to an increase of Cold War tensions just as it was a reflection of that Cold War. United States relations with its allies were influenced, most notably with Britain, France, and Belgium, as were contacts with such states as India and South Africa. The apparent success of the acquisition program built unrealistic visions of a *pax atomica Americana*, the dispelling of which provoked overreactions which only further excited a Washington atmosphere already heated by the debate over civilian versus military control of the atom. The pocketbooks of Americans and foreigners alike were affected, for the total expended directly and indirectly in dollars and pounds sterling was huge. It surely is one of the more startling examples of dollar diplomacy. Extravagant in scope, imagination, and expense, the campaign was conducted in great secrecy, with the right hand often uninformed of the activities of the left. In this case parliaments, congresses, chambers, cabinets, and state departments were only partially informed, while soldiers and businessmen made significant commitments affecting political arrangements.

In retrospect, the search for monopoly seems a wild-goose chase. One of the lawyers involved, John Lansdale, Jr., commented to the author that "Nobody had the faintest idea of how plentiful uranium was." Had they known, he went on, General Leslie Groves and his men would certainly not have tried to make deals with Sweden or Brazil. Yet until they knew, the endeavor seemed desperately necessary. If the situation was such that one nation could establish hegemony over the greatest destructive and constructive power source the world has known, then it was better to be that nation rather than not. The stakes were huge and potentially long-lasting; if it presently may seem that the world is better off to have several nuclear powers than only one, the ultimate result still remains to be seen.

An interesting aspect of this issue of such grave import for the balance of power among the great nations is the crucial role of Belgium. Though one of the smallest of nations, Belgium proved to have statesmen of great heart and steadiness of commitment to the Western, democratic cause. To several American participants, it is this small number of Belgians, and especially Edgar Sengier and Paul-Henri Spaak, who are the heroes of the story and of the protection of the West at the time when Europe lay vulnerable to Stalin's divisions. Of course there were many more individuals involved. Yet it is worth noting that because of the dense secrecy and haste with which the search was undertaken, it was more the product of individuals than of institutions, of personal efforts than of established procedures.

An account of Western efforts to procure uranium is difficult to construct. Only a select number of men were privy to the innermost dealings. All of them have been circumspect in their public comments, and some no doubt have carried important information to the grave. Operations were compartmentalized, so that few individuals ever had a grasp of the entire picture. The nations involved are chary about releasing documents and have applied restrictions well in excess of those normally prescribed by law.

The British and Americans were influenced in their efforts to obtain uranium by what they believed and suspected to be the tactics of their opponents. The extent to which the two governments were correct in their judgments or acted inappropriately cannot be known without access to files that will not be available to Western researchers for some while, if ever. Many German records for the war years are lost or destroyed, and the prospect of Soviet archives being soon opened is remote.

Moreover, the story has many strands. Much has appeared on the scientific developments, their technical nature and their political and national security implications. The detailed two-volume official *History of the United States Atomic Energy Commission*, based on the still classified documents of that Commission, is invaluable for an understanding of the American situation in this regard. Margaret Gowing's three volumes on Britain and atomic energy from 1939 to 1952 play a similar role for the British side. But there were other strands, nor can official histories cover all aspects equally well. The role of espionage is still obscure; the activities of private entrepreneurs and business firms cannot be easily traced, not to mention the feverish searches of individual prospectors.

Economy of space and clarity of focus require that attention be granted here only to the diplomatic give and take, with merely passing reference to related topics such as scientific debates or complexities of domestic politics. The existence of these should of course be kept in mind; they play an important role in the history of the period, if not in this account.

Archival restrictions currently prevent exploration of the topic past the early fifties, but absence of such extension does not cause too great a loss of understanding of the diplomatic effort. By then, the parameters of the situation had changed greatly. It was the assumed rarity of uranium, the secrecy surrounding the super weapon fueled by it, and the presence of war or the potentiality of war that lent such import to the negotiations in the 1940s. Slowly a number of these factors fell away. Germany was defeated, and the existence of an atomic weapon was terrifyingly revealed at Hiroshima and Nagasaki. In August 1949 Russia demonstrated her ability to produce an atomic weapon, and in October 1952 the British did likewise. Shortly thereafter the United States tested its first hydrogen bomb, a weapon requiring less uranium than earlier atomic bombs, and unanticipated reserves of the rare fuel were discovered.

Uranium-related diplomacy continued, but it took on a less frenzied tone as supplies became more readily available; mass production of bombs shifted emphasis from their manufacture to delivery systems, and authorities in the United States considered it possible—and a necessity—to turn over the development of uranium resources to private entrepreneurs. Concern remained high for secrecy of technology, limitation of the spread of atomic weapons, and the exchange of information regarding commercial and industrial usage of atomic energy, yet the scramble for an Anglo-American duopoly of uranium would be over.

My efforts to trace the campaign for rare ores have been aided by several persons to whom I owe sincere thanks. Mr. Edward Reese of the Modern Military Records Division of the United States National Archives was helpful. Ambassador Margaret Tibbetts, Mr. John Lansdale, Jr., Mr. Joseph Volpe, and Dr. Thomas Johnson shared their insights and experiences with me. Mr. and Mrs. David S. Clark, Mr. and Mrs. Kevin Miller, and Mr. and Mrs. Robert B. Stevenson offered shelter hospitality, and support. The Penrose Fund of the American Philosophical Society provided a welcome grant. Allegheny College kindly extended a sabbatic leave and a typing grant. Miss Margaret Moser, Miss Dorothy J. Smith, and Mr. Donald Vrabel at Pelletier Library gave able assistance. Professor Gregg Herken's suggestions were on the mark, and Ms. Gail Ullman and Miss R. Miriam Brokaw at the Princeton Press gave skilled guidance. The Belgian Académie royale des sciences d'outre-mer/Koninklijke Academie voor Overzeese Wetenschappen granted permission for a chapter published in one of its volumes to be reused here. Above all, Professor Jean Stengers of the Free University of Brussels, in this instance as in many others, offered significant encouragement.

My gratitude to my family is for even more than the happy atmosphere which enabled me to pursue this inquiry with support and enthusiasm. My parents, Professor Ernst and Dr. Louise Roberts Helmreich, each read all or part of the chapters and by their questions helped me to clarify the exposition. Nancy Helmreich typed numerous drafts as well as bringing sunshine to the darkest of winter days. The children, by their presence, hopes, and inquiries, provided animate reason for endeavoring to learn more about the roots of the atomic arms race.

To all of the above, I am indebted. Responsibility for any errors and shortcomings of the present account, however, falls only to myself.

Allegheny College
Meadville, PA

Gathering Rare Ores

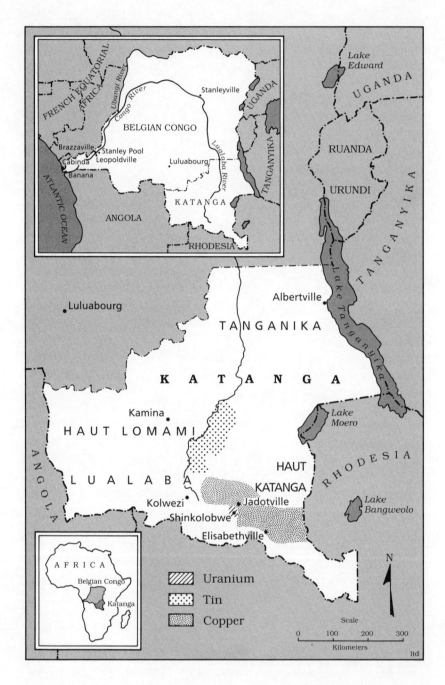

Province of Katanga, 1950

1 · Discovering the Need

In the summer of 1939 uranium was not an item which made much impression upon international affairs, world trade, or the public in general. Radium, with which it is usually associated when mined, was far better known for its use in scientific research and medical facilities. The ceramics industry did employ uranium to produce red and orange hues in its products, yet only about 100 tons of the metal were consumed the world over each year. Nearly 80 percent of this came from a mine at Shinkolobwe in Katanga Province of the Belgian Congo. When the threat of war increased the demand for copper and tungsten steel, even that mine was closed by its proprietors so that resources could be concentrated on exploitation of neighboring copper and cobalt mines.

This was to change, as did so much else, in the next few months. Scientific discoveries began to suggest the possibility of creation of a new and terribly powerful weapon from uranium. The outbreak of war lent an urgency both to the researchers' investigations and to the quest for the rare metal without which they could make little practical progress.

The desire of Great Britain and the United States in the nineteen forties and early fifties to gain control of foreign deposits of uranium and thorium which could provide the fuel for explosive atomic devices necessitated a major diplomatic campaign. The stakes were high and required unusual effort that nevertheless had to be sheltered from the public eye. Secrecy was essential both because of the danger of leakage of significant information and because the issues and tactics employed were not readily understandable. Diplomacy was carried out by an unusual mixture of military, civilian, and diplomatic personnel. Its twists and turns proceeded from technological discoveries and espionage as well as from a variety of conflicts: those of great powers as enemies, as differing allies, and as negotiators with independent and sometimes wary small powers. The dual goals were to assure the United States and Britain of a sufficient supply of uranium for their own weapons program and to deny their enemies access to the same material in a manner thorough enough to hamper or even forestall their atomic programs. Success was achieved on the first count, though at times the British feared their own research would be starved by the huge American appetite for radioactive ores. On the second count, success was more limited, though Russian progress (more so than the German) was no doubt slowed and made more expensive.

The quest for uranium reflected and furthered the transition of the role of major enemy of Britain and the United States from Germany to the Soviet Union after World War II. Its story therefore sheds light on a key source of tension and on the posture of the Western allies in the early years of the Cold War. Indeed, the American-British attempt to gain a preclusive duopoly on uranium regardless of whatever proposals for international cooperation were laid before the United Nations may arguably be among the prime origins of the Cold War, at least to the extent the Russians knew of the Anglo-American effort through their intelligence system.

The Early Search: Leslie Groves and Edgar Sengier

Recognition came slowly in the United States that uranium might be valuable to a weapons program. At the end of 1938 an isolated experiment in Berlin had suggested that uranium was fissionable. This was corroborated by further experiments. By the close of 1939 researchers at Columbia University in New York were investigating the possibility of creating a limited chain reaction, that is, of continuing fission once it had been initiated. Among them were the Italian scientist Enrico Fermi and the Hungarian Leo Szilard. Fear that the Nazis might be the first to achieve an explosive device utilizing the immense force of fission led Szilard to press for support from the U.S. government. Concerned also that the Germans might gain access to the best known source of uranium in the Belgian Congo, he and fellow Hungarian physicist Eugene Wigner approached Albert Einstein, then in residence in Princeton. As Einstein knew the Belgian royal family, they thought him the right person to warn the Belgians. Einstein agreed to write a letter, though to a lower-ranking individual. First a copy would be sent to the U.S. department of state. Were no response received, the letter would then be sent in two weeks to the Belgians.[1]

No prompt reply came from the state department, but the letter to Belgium was not mailed. Szilard in conversation with Alexander Sachs, an economist with access to President Franklin D. Roosevelt, learned that if government support were to be won the somewhat pessimistic views of Enrico Fermi and others at Columbia regarding the possibility of a fission bomb had to be overcome. Einstein again agreed to sign a letter, this one to Roosevelt. Drafted by Szilard and Sachs, it indicated that propagation of a chain reaction was imminent. It also warned that a frighteningly powerful bomb might conceivably be constructed, urged government awareness and support of research activities, and noted that the Congo was the best source of uranium and that Germany had stopped sale of Czech uranium.

After meeting with Sachs on 11 October 1939, Roosevelt appointed an Advisory Committee on Uranium. The need for defense research in the still neutral United States became more clear by June 1940, when the National Defense Research Committee (NDRC) was created under the leadership of Dr. Vannevar Bush. Bush, an inventive applied mathematician and electrical engineer, was a former vice-president of the Massachusetts Institute of Technology and currently head of the Carnegie Institution. The NDRC supervised the Uranium Committee until November 1941, when that committee's work became so important that it was placed directly under the Office of Scientific Research and Development, the governmental agency which also oversaw the National Defense Research Committee.

Fission research accelerated. In June 1942 the U.S. Army formed an engineer district—eventually named the Manhattan Engineer District (MED)—to assist with the effort. Among its chief responsibilities would be construction of production plants. It made little progress until in mid-September a newly promoted brigadier general, Leslie R. Groves, was appointed its head. Former deputy chief of construction in the Corps of Engineers, Groves was able, hard-driving, contentious, and blunt. He had a record of getting things done. Smart, he did not mind letting others be misled into thinking he was not. Thorough in preparation, he was nevertheless known to take decisions and actions abruptly. Groves had a strong sense of intuition about people and would follow it with a remarkably high degree of success, as one colleague noted.[2] In the Manhattan Engineering District he would develop a policy of compartmentalization of activities and knowledge so that few persons other than himself knew all that was going on. For security reasons also, he would work with as few people as possible.

On 17 September 1942, the day Groves was informed of his new duties, he discussed with an assistant the lack of uranium. There were few developed uranium mining sites in the Western hemisphere, for the market for uranium and most of the metals with which it is commonly found had simply not warranted their discovery or exploitation. In 1939 uranium imported to the United States brought only about 83 cents per pound, hardly enough to encourage much prospecting. Some had been found along with radium in Colorado, but its mining there had dropped off sharply after richer deposits discovered in the Congo were mined at far lower costs. The Eldorado company in Canada did produce uranium as a by-product of its gold mining ventures, but the amount was limited.[3]

Groves quickly learned that the only available supply of uranium was 1,200 tons of high-grade (65 percent) ore of which the MED had learned almost by accident ten days earlier. It was stored on Staten Island by the African Metals Corporation, an affiliate of Union Minière du Haut Katanga.

Union Minière (UM) was a partial affiliate of the Société Générale de Belgique (SGB), one of the two or three largest investment concerns in Europe and one of the oldest. Stock in UM was also held by individuals and holding firms in a number of countries, and stock in some of these investment firms was in turn held by the Société Générale. Belgium corporate law had no "arm-length" provisions, and interlocking directorates were common. The connection between UM and SGB has traditionally been close. In 1981, well after UM had been nationalized by Zaire in 1967, the company would become a wholly owned subsidiary of SGB. In the 1930s Union Minière was deeply involved in the discovery and development of non-ferrous metals in Upper Katanga, in the southwest section of the huge African country first established as the Congo Free State by Belgian King Leopold II in 1885 and eventually turned over to Belgium as a colony in 1908.

The uranium which interested Groves had been brought to New York on the order of Edgar Sengier, the managing director of Union Minière. Enterprising and far-sighted, he had encouraged vast investments by his firm in developing the initially unprofitable copper mines of Katanga. Although concentrating on copper, Sengier became aware that uranium might be of greater value than most people thought when he visited a fellow Union Minière director, Lord Stonehaven, in England in May 1939. Lord Stonehaven arranged for Sengier to meet Sir Henry Tizard, director of the Imperial College of Science and Technology, and deeply involved in British defense research.

Tizard knew that an outside possibility existed that a powerful bomb might be created from uranium, but he judged that possibility as only one in a hundred thousand. Nevertheless, alerted by a British researcher of the desirability of denying uranium to Germany and securing it for Britain, he asked of the Union Minière officials that his government be granted option to purchase all radium-uranium ore produced by the UM mine at Shinkolobwe in Katanga. Tizard did not press the matter vigorously, for he doubted the utility of uranium. Sengier refused the proposition, perhaps because the price the Belgian believed his ore was worth had not been offered. He did agree to let the scientist know of any abnormal demand. In parting, Tizard warned that the Belgian held "something which may mean a catastrophe to your country and mine if this material were to fall in the hands of a possible enemy."[4]

Sengier was next approached a few days later by French scientists led by Frederic Joliot-Curie. The latter was the son-in-law of the famous scientists and co-winner with his wife in 1935 of a Nobel Prize for work in nuclear physics and radiation. Would Sengier join them in an effort to explode a uranium fusion bomb in the Sahara? Sengier accepted the proposal in principle and agreed to provide the necessary ore. Outbreak of war and the invasion of France, however, prevented any development of these plans.

The possibility that Belgium would be invaded and communication between Brussels and the Congo cut off led the directors of Union Minière to take precautions. Sengier was quietly sent to New York in September 1939, a few weeks prior to his sixtieth birthday, with full powers to conduct the firm's business should contacts with its European directors be broken. Before leaving Brussels, Sengier ordered existing supplies of radium in Belgium and uranium ore at the refining plant at Oolen, Belgium, near Antwerp to be shipped to Britain and the United States. The radium arrived, but shipment of the Oolen ores was delayed; they fell into German hands in June 1940 when Belgium was overrun. Some 3,500 tons of uranium compounds, some already partially processed, thus became available—barring bureaucratic hold-ups—to the German atomic research program.[5] Stockpiled ore in the Congo was shipped to New York promptly; it was this ore which eventually attracted the attention of the MED.

To Sengier's surprise, American officials at first took little interest in the uranium. In March 1942 Sengier talked with Thomas K. Finletter and Herbert Feis of the state department. These economic experts were more interested in cobalt than uranium, even after Sengier suggested that the latter was more important. Twice in April he raised the matter without significant response. This was no doubt because the state department did not know about the current U.S. research activities and indeed would not be informed until just before the Yalta conference of February 1945—such was the secrecy surrounding the project and so closely did Roosevelt hold his cards to his vest. Yet, as General Groves points out in his memoirs, the connection of valuable radium with uranium was widely known, and there were enough articles on recent research in the press to suggest the importance of uranium on its own.[6]

Apparently the Executive Committee of the Advisory Committee on Uranium (now known as the S-1 Committee) chaired by Dr. James B. Conant, the chemist president of Harvard University, also saw no need to acquire any additional supplies of uranium. At least that was the position taken at its 9 July 1942 meeting. It was expected that sufficient uranium might be obtained via the Canadian firm of Eldorado Gold Mines, Ltd., from which two small (6 to 8 and 5 tons) orders had been purchased in 1941. Eldorado's old mine on Great Bear Lake would, however, have to be drained and brought into repair. In order to persuade the company to take these steps, an order for 60 tons of oxide, the least amount necessary to make the re-opening economically possible, had been placed. But then in August the S-1 Committee learned that Boris Pregel, a White Russian who in 1917 fled to France and established various connections including that of a sales agent for Union Minière and also for Eldorado, was attempting to purchase 500 tons of Sengier's ore.[7] On 11 September, Bush, as head of the National Defense Research Committee, suggested to the army the imposition of export controls on uranium.

In a marked alteration of its July posture, the S-1 Executive Committee recommended the imposition of export controls and the purchase of Sengier's ore. It was pushed to this point by the realization that otherwise it might, because of Pregel's connection with Eldorado, be buying ore for shipment to the United States which was not mined near the Arctic Circle but already on an island in the middle of New York Harbor. Also, Sengier's hand-sorted ore averaged 65 percent uranium oxide, while Colorado and Canadian ores held only .2 percent. That this fact may have been the chief motive for the change in the attitude of the U.S. officials is indicated by the Committee's confidence that a sufficient supply of uranium oxide for the war effort was available and its acceptance of an army recommendation that the flooded mine at Shinkolobwe in the Congo not be reopened.[8]

Colonel K. D. Nichols, an experienced hand at the Manhattan Engineering District, discussed all this with Groves on 17 September 1942. Both were more concerned about the limited ore supplies than was the committee. They acted promptly. Nichols arranged export controls through the state department, and the next day he visited Sengier at his office in the Cunard Building. Once the Belgian was convinced that Nichols meant to deal seriously, matters moved quickly. Within an hour it was agreed that the United States would buy all the ore stored on Staten Island and that the Americans could have first option on the 1,000 tons stockpiled above ground in the Congo; these were to be shipped immediately. Contracts were worked out later and signed on 19 October. To ensure secrecy, correspondence was limited, and the Federal Reserve Bank, which oversees banking activities in the United States, was instructed not to mention the transactions in its reports.[9]

A Joint Anglo-American Effort

If in 1942 the S-1 Committee was concerned only with procuring uranium for the war effort, by the following year this viewpoint had changed. Research had progressed to the point that creation of a bomb seemed a possibility, although the time frame was still vague. The implications this held for the future balance of world power were great. Groves in particular was concerned that at the war's end North American supplies of uranium might be exhausted and the United States would have no control over the world's best source in the Congo. Those ores were especially needed because of the physical and chemical properties of uranium and the inefficiencies of the technology then available to separate the small amounts of fissionable isotopes in the element.[10]

Vigilant that he not be left to work solely with low-grade tailings from Colorado and poor Canadian pitchblende, troublesome to mine and difficult to refine, Groves rushed a message to Roosevelt in August 1943. The president was meeting in Quebec with British Prime Minister Winston Churchill, and Groves knew that the agenda included the patching up of Anglo-American differences on interchange of atomic research information.

Exchange of war research information, especially regarding short-range radio waves, had begun near the end of 1940. By the end of 1941 uranium (S-1 or Tube Alloys, the British code word) was being discussed, and the possibility of a joint production plant in Canada was mentioned. But little progress was made during the following year. American science administrators were reluctant to share information which went beyond the prosecution of the immediate war or into areas that the British themselves were not investigating. Bush and Conant both suspected that the British were looking toward postwar commercial advantages. The British insisted that they were concerned only for postwar security.

On 22 July 1943 a compromise formula had been outlined by Churchill at a meeting in London with American officials. In more polished form, this became the Collaboration Agreement which Churchill and Roosevelt would sign at Quebec on 19 August 1943. Among other points it called for full scientific research and development exchange to occur among persons working in similar endeavors; interchange on construction of large-scale plants would be by *ad hoc* arrangements. Neither party would use the bomb against the other, nor would either use it against a third party without mutual agreement. To facilitate collaboration, a Combined Policy Committee was to be created under the chairmanship of U.S. Secretary of War Henry L. Stimson. Of the six members, three were to be from the United States, two from Britain, and one from Canada.

The reestablishment of better relations with the British on uranium research was fortunate for Groves. Concerned that "in the future the United States would not lack the essential raw materials . . . suitable for the production of atomic energy," he had employed in May the Union Carbide and Carbon Company to assess world uranium supplies.[11] Two days after the Quebec Agreement was signed, the Military Policy Committee approved Groves' recommendation that the "United States allow nothing to stand in the way of achieving as complete control as possible of world uranium supplies."[12]

If the Americans were upset at British desires to look beyond the war years, they were interested in doing so themselves. That fall Groves sent Captain Phillip L. Merritt, a geologist, to the Congo to search for other uranium sources there in addition to Shinkolobwe. Merritt found none but learned that, because the ores at Shinkolobwe had been hand sorted and only the richest were exported, the tailing dumps contained supplies varying in uranium content from 3 to 20 percent. These were promptly purchased, and efforts were made to persuade Sengier to re-open the mine and to promise its full output to the United States. The Belgian, aware of the importance of such an action and hesitant to take this step on his own, refused to give an option or first refusal on all ores produced; he did continue negotiations regarding sale of specific tonnages of newly mined oxide to the United States. These did not progress to fruition because of disagreement on price.[13]

Unable to get the full control it desired, the American Military Policy Committee turned to the Combined Policy Committee (CPC), established as a result of the Quebec accord. Moving the matter to this level promised two advantages. First, British involvement would bring into play both the long-standing close relationship between Britain and Belgium and the implied concerns of the British shareholders, who held about 30 percent of the Union Minière stock. Second, it would provide some cover for Sengier as an individual company officer if the cooperation of the Belgian government in exile could be elicited. As Groves later explained the matter:

> The Military Policy Committee felt that if Great Britain took advantage of the location in London of the Belgian Government in exile or, later, of the normal British strong influence over Belgian policy, it could and would secure a monopoly over the Belgian Congo raw material. The United States would then be in a most disadvantageous position, even if it should take the British a number of years to develop atomic energy either for military or commercial purposes. [Although the Manhattan Engineering District was exploring a contract for reopening the mines] . . . it was realized that our best prospect for obtaining an exclusive long-term commitment from the Belgians would be through the medium of a governmental agreement between Belgium on the one hand and the United States and the United Kingdom on the other.[14]

The British by this time were themselves more interested in uranium. Their initial research needs had been small, in the range of less than one and a half tons of oxides through 1941. These had been obtained from Oolen at the urging of British scientists and with the aid of John Anderson, then Lord Privy Seal. Two more tons were bought from Canada at the end of 1941 through the Canadian National Research Council so as to prevent the Eldorado company from learning of British interest and driving up the price. In 1940 a uranium subcommittee of the British Committee for the Scientific Survey of Air Warfare had been established. Known by the cover name of the MAUD Committee, it investigated the possibility of creating an atomic bomb. In July 1941 it issued two reports on the use of uranium for a bomb and as a source of power, estimating that a bomb could be produced by the close of 1943.

The reports were read by Sir John Anderson, currently Lord President and soon to be chancellor of the exchequer. A former governor of Bengal, experienced as permanent under secretary in the home office, a member of parliament, and a current director of Imperial Chemical Industries, he was Churchill's specialist on uranium. Britain was already involved in a program of preclusive buying of scarce and valuable commodities from neutral countries. Anderson determined to further steps already underway to prevent Nazi access to uranium reserves and also to avoid the driving of prices upward by speculation. At his instigation, the British government bought up various small concessions, including the Urgeirica mine in Portugal. These were not expected to produce great quantities, and the purchases were considered primarily preemptive. The potential of Eldorado in Canada was far greater; Anderson, supported by the treasury and the Department of Scientific and Industrial Research, moved to gain control of the company.

Emissaries were sent to meet with C. D. Howe, Canadian minister of munitions, and with the president of the Canadian Research Council, Dr. Chalmers J. Mackenzie. Howe agreed that control should be obtained, but he was reluctant to use governmental wartime powers to do so because of the attendant publicity. Howe suggested that it would be better for the Canadian government quietly to buy a majority share of the company, a move which could be arranged by his friend Gilbert LaBine, who Howe believed controlled as well as managed the company. Bush was contacted in Washington and indicated his approval. A million shares were purchased from LaBine before it was discovered that, in order for the government to gain a majority of the stock, another million would have to be purchased on the open market. The cost of the effort was rising.

There seemed no need to hurry; to do so, in fact, would invite speculation. An order for 20 tons of oxide was placed and no effort made to inform Eldorado of future needs. Then the British learned indirectly that Groves, through a series of orders negotiated in 1942 and the following winter for some 850 tons of ore, had essentially taken over the entire Eldorado production for the next three years. Moreover, he had also arranged for Eldorado's refinery to process the rich Belgian ores, giving them priority over Canadian ores. The refinery's capacity was limited and thus preempted for what turned out to be five full years.[15]

How the British and Canadians allowed this to happen never became truly clear; apparently it was a case of inattention, overburdened men, poor communication, and perhaps insouciant capitalism. The machinations of Pregel were also involved; already disliked by the British, he soon became *persona non grata* with Groves and would be cut out of later dealings. The affair naturally led to hard feelings between the British and Americans. Groves' action was scarcely tactful, especially after the British had so courteously tipped their hand to Bush. His act reflected both the poor Anglo-American scientific relations that winter of 1942-1943 and the importance of the reconciliation achieved at Quebec the following August.

The incident also created friction between the British and Canadians. In the summer of 1943, when the British asked Canada for 120 tons of uranium that she could not deliver, Mackenzie visited Groves in New York, hoping to reach some arrangement which would allow the Canadians access to some of their own uranium. Groves was reluctant to part with the amount of oxide Mackenzie requested but did hold out the hope that mine production and refinery efficiency might be improved to meet this need. Investigations were launched, but it was clear that no solution would be rapidly forthcoming. The Canadians resented the pressure brought to bear on them by London in the name of old relationships: one British representative noted a Canadian feeling "that some quite artificial and non-realisable condition is being forcibly substituted, by the British, for the obvious and natural state of complete cooperation with, and dependence on, the Americans."[16]

Facing American control of the Eldorado production, Colonel J. J. Liewellin, a British member of the new Combined Policy Committee, called for an independent British approach to the Belgians. Anderson argued against this, fearing that a separate approach might encourage the Belgians to play the British and Americans off against each other to raise prices. Anderson preferred a joint approach, which especially made sense if the new-found harmony of Quebec were not to be disrupted. He had also heard through informal channels that Sengier preferred not to be put in the position of choosing between the two powers—a choice that later might be subject to criticism by his company's board of directors—but might be amenable to a joint approach.[17]

The American members of the CPC were still hoping that Groves would reach his own agreement with Sengier when the British informally mentioned the possibility of talking with the Belgian government. Somewhat later, at the December 1943 meeting of the committee, when the topic was formally raised, the Americans responded more eagerly. They were no doubt pleased that it was the British who had been maneuvered into taking the first step.

On 17 December Major General Wilhelm D. Styer, chairman of the subcommittee dealing with uranium, told the Combined Policy Committee that control of the Belgian Congo supply was of "great importance" and required action at the highest levels.[18] Groves reported this to Roosevelt, and on 15 February Bush received the president's permission to proceed. Thus it was the caution of Sengier, whom Groves always considered a staunch ally and highly cooperative, fear of British domination of Belgian policy, and determination to gain long-term control of the ore supply that led the Americans to treat the matter more as a political than a commercial transaction and to seek a joint Anglo-American approach to the Belgian government. The early suspicions of British postwar commercial considerations had not entirely died away; these may also have prodded the Americans, who would have preferred to go it alone, into co-opting the British through fear of being left out. From the British side it was also fear of being left without supplies, joined with concern that they might be trapped into an expensive bidding competition, that recommended a cooperative approach.

2 · The Cornerstone: Agreement with Belgium

Initial Contacts

The decision formally to approach the Belgian government in exile was taken at a meeting of the Combined Policy Committee on 17 February 1944. Among those present were U.S. Secretary of War Henry L. Stimson, Bush, Conant, and from Britain, Field Marshal Sir John Dill and Sir Ronald Campbell. Harvey Bundy, an American lawyer, and an Englishman, W. L. Webster, shared the secretarial responsibilities. Negotiations were to be carried on by the American ambassador in London, John G. Winant, and by Sir John Anderson, British chancellor of the exchequer, chief overseer of uranium policy for his country and possessor of close contacts with the Belgians. Anderson was expected to play the leading role. General Groves was reluctant to bring in a state department official at this point. Atomic energy matters had thus far been handled through the war department. Using the same people meant less security risk and firmer control for Groves. He therefore appointed Major Harry S. Traynor, then working at the Oak Ridge atom facility, to carry instructions to Winant and serve as liaison with himself.[1] Traynor arrived in London Saturday, 18 March 1944, and that weekend met with Winant at length. He explained the need for a long-term treaty and for the creation of a commercial entity to handle the anticipated uranium purchases.

At its 17 February meeting the Combined Policy Committee had drafted a tentative proposal for an Anglo-Canadian-United States corporation operating under the Combined Policy Committee. Its purpose would be to purchase or gain control of uranium outside the bounds of the United States, the United Kingdom, and the British Commonwealth. Control of uranium within the territories of the participating countries was to be left directly to those countries. The proposed preamble stated that, as adequate supplies of uranium were vital to those countries, it would be unwise to exhaust their reserves without taking steps regarding the richest known deposits, which were in the control of other nations. Continuity was necessary, and as no American administration could bind its successor without a vote by congress, which would create unwanted publicity, some sort of private corporation was envisaged. United States law, however, prevented the transactions of a corporate firm from being kept secret. At British suggestion, the concept of a Combined Development Trust was adopted. When on 13 June an Agreement and Declaration of Trust was jointly signed, the Canadians were not participants. They had not been party to the Quebec Agreement and therefore were dropped from the Trust proposal for the sake of consonance. As in the Combined Policy Committee, however, they would have one vote in the Trust. Research results achieved since February brought another change, as the acquisition of thorium as well as uranium was brought under the Trust's purview.

Winant and Traynor met with Anderson and W. L. Gorell Barnes, representing the British War Cabinet. The chancellor suggested that the first approach to the Belgians be made through Camille Gutt, the experienced Belgian minister of finance who had many years of contact with the British. Anderson would tell Gutt in strict confidence of the experimental atomic work and indicate there was no certainty of what the results might be but they could be of "greatest significance."[2] If they materialized, it would be important that uranium not fall into the wrong hands and that Britain and America obtain control of as much as possible of the world's supply. While his mind was open regarding the nature of the agreement, Anderson thought the first objective should be an option for a number of years on all the uranium in the Congo. Price would be a difficult issue; perhaps agreement could be reached on the basis of the metal's market value over the past few years.

Winant approved the approach: both men saw need for quick action. Anderson invited Gutt to his office late that afternoon of 22 March 1944 and asked his view of the best means of reaching an arrangement. Although Gutt would later comment that he found Anderson pompous, the Belgian proved sympathetic. He assured Anderson that his government would cooperate to prevent the material from falling into the wrong hands. Gutt bristled at the word "control," and Anderson quickly interjected that he was not thinking of anything which would derogate from the national sovereignty of Belgium. The powers just wanted control of the product, not of the mine, and since the matter was of such importance it could not be looked on as just a commercial transaction.

Indeed, it was of such importance it might be dealt with in any ultimate peace treaty.[3] Gutt said he would have to consult with Prime Minister Hubert Pierlot, Foreign Minister Paul-Henri Spaak, and Minister of Colonies Albert de Vleeschauwer but saw no need to include the governor general of the Congo in the talks; after all, the matter was solely up to the Belgian government.

The Belgian cabinet met that evening and promptly agreed on an affirmative response in principle on two conditions. First, Belgium was to be kept informed "of all the developments the Anglo-American research had and would produce (save those which were strictly military secrets)."[4] Second, price and delivery terms should be satisfying to Union Minière.

Before a joint meeting of the representatives of the three powers on 27 March, Anderson, Winant, Traynor, and Gorell Barnes had a quick huddle. Winant cautioned against including any mention of control of uranium in a peace treaty, for fear this might tend to postpone any final agreement. Control of uranium was needed now, without reference to later review. He assured the British that the United States had no constitutional problem in entering an arrangement continuing past the end of the war, as the president's authority to defend the country was a continuing one. Anderson in turn warned that Gutt was uneasy over the word "control"; it was clear that the Belgian ministers would need careful handling.[5] Anderson would therefore limit himself to getting their reaction and to discovering the potential output from the Congo.

The meeting with Gutt, Spaak, and De Vleeschauwer went well. Gutt reiterated his country's desire to assure that uranium did not get in the wrong hands and to collaborate with the two powers. To Anderson's inquiry regarding the amount of ore which might be available, De Vleeschauwer pleaded ignorance and insisted that Sengier be brought over from New York. There was mutual agreement that to preserve secrecy no information would be sent to the Congo. The Americans, informed by their specialists, knew more about potential output than did the Belgian ministers. More puzzling is that, while aware of exports to the United States in recent years, the Belgians were not apprised of Sengier's contracts or of his current negotiations with Groves.[6]

Brigadier General Groves and his colleagues in the Manhattan District "were delighted" that Sengier would be involved, "since his presence during the negotiations would be an insurance against delay and unfortunate questions."[7] But the American general underestimated Belgian bargaining capacity. On 8 May, at a meeting of the representatives of the three powers, Gutt reaffirmed his desire to cooperate; he then added that, if the research undertaken with Belgian materials brought, in addition to military possibilities, results capable of commercial and industrial exploitation, the Belgian government would expect to share in those benefits. Anderson quickly pointed out that the Congo was not the only source of uranium and that thorium was also being investigated. Gutt admitted that Belgium should share only to the extent and as long as the material she supplied was instrumental in securing the results. Anderson and Winant saw little difficulty in working out an "equitable adjustment of interests" if the work underway produced results capable of commercial exploitation.[8]

When Sengier joined the meeting a few minutes later, he reported that at Shinkolobwe uranium had been extracted to a depth of 79 meters. Test probes suggested there were 101 more meters of good ore which would yield 10,000 tons of 50 percent to 60 percent uranium oxide (U_3O_8). This was the estimate of both the company and engineers sent by the U.S. Army the preceding fall. If the mine were to be reopened, equipment would have to be transferred from a nearby copper mine, new equipment obtained, electricity provided to pump the mine dry, and workmen assembled. As a private citizen, Sengier had been reluctant to dispose of the mine's ore because, in view of its importance, he felt this decision of national import exceeded his powers. Given the urging received from the Belgian government he could now accept the responsibility; the labor supply issue could be overcome. He then revealed his previous negotiations with the U.S. Army and handed Anderson a letter of intent and a draft contract between the Manhattan District and the African Metals Corporation which he had received some weeks earlier. These had been drawn up as a result of the September 1943 American mission to the Congo.[9]

The contract called for delivery of 3,400,000 pounds of oxide at $1.71 per pound during a period of 18 months following the first production of high-grade concentrates. This date in turn would be about 18 to 20 months from the time orders were placed in the United States for the necessary equipment, specifications for which had been drawn up in Africa by the management of the mining company and the American team. The resumption of production had also been made conditional on the supply of additional workers. These basic terms had been agreed upon in March 1944, but since then the U.S. Army had suspended negotiations.[10]

In concluding, Sengier asked that Union Minière be allowed to honor its contracts supplying the world ceramics industry. He also insisted that his company not be discriminated against to the benefit of the Canadian Eldorado Company in the supplying of radium to Canada. Earlier, that country had set up a protective tariff on radium so steep that Union Minière was effectively excluded from the Canadian market. In retaliation Sengier had demanded return of radium that his company had lent to Canadian hospitals for medical purposes.

Another meeting was scheduled in three days. Meanwhile British and American representatives met privately with Sengier. For the British these were Gorell Barnes and Sir Thomas Barnes, the British treasury solicitor; for the Americans, Traynor, and Brigadier General E. C. Betts. The latter was judge advocate general of the European theater of operations assigned to Ambassador Winant by General Dwight D. Eisenhower to advise Winant on matters of war department interest lying outside the European theater. According to the report of their 9 May meeting, Sengier was willing to sign immediately a contract for 3,440,000 pounds of uranium along the lines of the earlier draft contract. He would make no commitment for supplies beyond this amount, however, unless an agreement were negotiated by the governments. A "right of first refusal" deal was acceptable to Sengier on the assumption the price paid would be the current market price at the time of sale. Differences arose on this point. How was an open market price to be determined? The English and Americans suggested that the price should be calculated from time to time on the basis of cost of production plus a reasonable profit margin. Sengier was not amenable to this and said he would advise his government against such an arrangement.[11]

While Sengier thought an allied mission working with company officials to survey further Congo resources might be possible, he "did not take kindly" to the notion of observers or technicians at his mines. This attitude was based on "an alleged fear" that the presence of "fat-salaried" Americans might cause unrest among local laborers.[12]

The British and American negotiators carefully avoided raising the issue of radium tariffs, for this was a matter of Canadian sovereignty. As for uranium supplies for the ceramics industry, Sengier believed these varied from 40 to 80 tons a year.

The negotiators were convinced that Sengier would use his considerable influence to dissuade the Belgian government from a long-term contract. They recommended that a contract for 3,440,000 pounds be signed at once and that further negotiations be pursued for a long-term deal. The allied representatives saw the question of whether the Belgians would allow a "first refusal" contract on reasonable terms as the "crucial point." They did not believe Sengier's attitude stemmed from a desire to protect his position vis-à-vis his stockholders by means of a government cover.

It seems to us that it is in the interests of the Belgian Government and of M. Sengier alike to do their best to avoid entering into any commitment which would effectively secure to us on reasonable terms any of the uranium in the Belgian Congo beyond the first 3,440,000 lbs. From the commercial point of view they probably anticipate that our experimental work will produce results which will mean that at a later date they will have a sellers market; and from the political point of view it must obviously be a paramount consideration for the Belgian Government to avoid if possible a commitment in a matter of this nature covering a long period after the liberation of their country.

In these circumstances we are of the opinion that the only way of securing a satisfactory long term arrangement will be to bring pressure to bear upon the Belgian Government in connection with the negotiations for the supply to them by our two Governments of their extensive requirements for Belgium on its liberation. Whether such pressure should be exerted on the present Belgian Government or whether we should be content with an agreement in vague terms pending the liberation of Belgium is essentially a political question on which we do not feel that it is in our province to advise.[13]

At the general meeting on 11 May, Chancellor Anderson proposed a three-phase undertaking. First, an immediate contract for 3,440,000 pounds of uranium oxide would replace the draft contract. Second, the Belgian government would ensure that further amounts needed for military requirements would be made available at a cost-of-production-plus-reasonable-profit price. Third, when requirements for military purposes were met and there was need for further supplies related to possible commercial and industrial exploitation, the terms would take into account the Belgian government's interest in an equitable share of the benefits of such exploitation—provided Belgian ore were used.[14]

Gutt insisted that the Belgian government should have means of assuring that material used in phase two which eventually led to processes capable of commercial and industrial exploitation came under phase three. As it could be argued that all phase-two work had eventual phase-three implications, Winant and Anderson maintained the matter was one of good faith; no resolution was achieved. Sengier, no doubt pleased to see his immediate contract separated from the long-range issue, volunteered to push ahead with preparations for resuming mine operations even before the final contract was drafted. He was optimistic that by the time the mine had reopened, Belgium would be liberated and skilled labor could be taken to the Congo. It was agreed that Sengier would return to New York, where details for his contract would be worked out. Meanwhile in London, Anderson and Winant would draft a longer-range agreement.

This they did, and within a week Winant carried a draft of the long-term arrangement to Washington for review. Article 1 stated that the "Two Governments shall have the exclusive right to purchase all uranium and thorium ores in the Belgian Congo. . . ."[15] Anderson in the meantime had realized that this wording, though giving complete control, "might morally commit us to buy all material even though in the future we did not have need for it."[16] A "first refusal" phrasing was therefore incorporated in the third article, which finally stated that Belgium would take the necessary steps to assure that no other government or persons would acquire rights to the ore unless the two governments had refused the offer of ore; the Union Minière's existing contracts with the ceramics industries would be honored. Such wording was also considered in better keeping with the 11 May discussion. Price, after delivery of the first 3,440,000 pounds, would be on a production-cost-plus-reasonable-profit basis. Cost of ore purchases beyond "military and strategic" needs would be determined from time to time, taking into account Belgium's desire for equitable participation in commercial benefits which might accrue.[17]

Belgian Counterproposals

The Belgians took some time in reviewing the proposal, and in the middle of July they submitted their own. By then Sengier had discussed contract terms with Sir Charles Hambro and Groves and submitted a tentative basic accord; the allies had landed at Normandy; Caen had been freed; and the Belgian ministers looked forward to the liberation of their own country. They asked that Belgian ceramics industries in particular have prompt access to the uranium oxide, if this did not harm the war effort. More troubling were three further positions taken by the Belgians. First, there should be no protective tariff in the United States or United Kingdom against Congo radium. Second, after Belgium's liberation, ores in excess of military requirements could be used in Belgian industries under the control of the Belgian government. Third, for the postwar period, "the Belgian Government would agree only to declare itself ready to enter into discussions, with a view to the acquisition and common utilization by the Three Governments of the uranium and thorium produced in the Congo."[18]

Stimson and Hambro happened to be in the United Kingdom at the time the Belgian response was received. The secretary of war took the position that if the Belgians would make no commitment regarding the postwar period at present, the two powers should unilaterally declare their intent to insist on such an arrangement later. This would then provide a basis for easier political action if it were necessary. The British liked the firmness of this position, and it was agreed at a Sunday evening meeting on 16 July to push for a postwar commitment.[19]

When informed of the Belgian counterproposals, the newly functioning Combined Development Trust, meeting in the United States with Groves and Sir Charles Hambro the key American and British members, balked and referred the problem to Anderson and Winant. These two met on 19 July with Barnes, Betts, and R. S. Sayers, director of the British uranium (Tube Alloys) project.

Anderson expressed sympathy for the Belgians' desire to use limited quantities of untreated ore in their own industries. Winant feared that the Belgians might in the future attempt a power project, expensive as it might be, rather than simply use the uranium in ceramics. The two eventually decided that the Belgians were referring only to routine uses; if interest were later expressed in new areas, these could be subject to review. Both men felt strongly the need for a long-term refusal clause, though Anderson feared that the Belgian ministers would plead they held authority only for the duration of the war. Winant argued for a 99-year control provision, saying that Secretary of War Stimson attached "high importance" to the inclusion of a binding clause rather than a vague promise to hold further discussions. General Betts saw such an agreement as providing the "best security for the Belgian Government ownership and control of these war potentials since the Two Governments would have under the agreement a very real interest in protecting their rights to these materials" which Belgium would not be in a position to protect. This was a view he was to reiterate to Bundy and others the following day; at that gathering it was agreed that the two governments could protect the asset against the evil-minded, while Belgium could not. They considered themselves entitled to control of the uranium because of expenditure of lives and treasure; "Belgium cannot stand aside and barter when Peace is involved."[20]

Anderson and Betts thought that the Belgians would buy this argument of great power protection only if they could later participate in the commercial benefits from exploitation of the oxide, and so phrasing to this effect was agreed upon. It was assumed that if the Belgian ministers were willing to make any arrangement for the postwar years they would accept a 99-year period as easily as 10 years.

The English and American negotiators also agreed that as what was contemplated was officially a technical rather than a political accord, no pledge could be given on tariffs. Nor could they promise to purchase all Congo radium and try to market it, as this could not be justified in terms of defense. The Trust should also not promise to indemnify the producers, for in negotiating with Sengier the Trust had settled on a price "so favourable to the producers as to make it fair that they should bear all the risks of marketing radium."[21]

A serious question remained. What if the Belgians refused to sign on grounds that the postwar years were beyond their current authority? Could they then be asked to accept a provision under which they promised to use "their best endeavors to secure a ratification" of commitments which they thought were beyond their legal power to make at this time?[22] This alternative wording was held in abeyance as a possible "fall-back" position. First the men would meet with the Belgians to discern their position and then submit the draft agreement with this further alteration if the conversations so dictated.

On 28 July Prime Minister Hubert Pierlot, Spaak, and De Vleeschauwer met the British-American team. After brief fencing, the notion of compensation in the price of uranium in lieu of alteration of the radium tariffs was accepted, subject to further Belgian review of the prices. Anderson then offered neutralized uranium (i.e., uranium after it had been processed for military purposes) "for existing contracts for ordinary industrial purposes."[23] Again the Belgians were amenable provided this was acceptable to the industries.

Then the real issue came into play. When Anderson asked for either a commitment of indefinite duration or for a specified long term, Spaak brought forth an argument that the Anglo-American team had overlooked. The General Act of the Berlin West African Conference of 1884-1885 (creating the Congo Free State), revised by the Brussels Conference of 1890 and the Convention of St. Germain of 1919, required the Congo to be a free-trade zone. Any agreement to restrict sales to two countries would conflict with obligations formally and seriously undertaken. Deviations could be made during the war years, but certainly not for postwar years. Chancellor Anderson protested that such an attitude was contrary to early promises of Belgian cooperation. In any case, would a long term contract to restricted destinations be a conflict? De Vleeschauwer thought it would. But what if the military importance of the uranium were established and known throughout the world? Could not the ministers say to their countrymen that an agreement with the two powers protected the national patrimony? The three Belgians acknowledged this possibility and said they would go as far as possible, subject to "over-riding international obligations."[24]

Anderson was satisfied with the results of the meeting. The Belgians seemed to be accepting the neutralized ore and had not mentioned reopening their treatment plant at Oolen. If they did so in the future, they could be granted ore which, when refined, would be exchanged for neutralized uranium. As far as the Congo free-trade issue was concerned, that argument had to be accepted; it seemed to be the only one raised against the long-term first refusal clause.[25]

The chancellor and his staff drafted a new version of the agreement and forwarded it to the Belgians with Winant's approval on the last day of July. The draft was more detailed than that of a few weeks earlier, and the articles were rearranged. Article 7 recognized Belgian desire for participation in commercial benefits and established a first-refusal right to the United Kingdom and the United States for 99 years, "subject to any existing international obligations of the Belgian government."[26]

The U.S. war department was not entirely pleased with Anderson's draft, however, and forwarded six proposed changes. The negotiators in London rejected four, accepted one, and agreed to suggest one to the Belgians. The war department's main objection had to do with the phrase citing the international obligations of the Congo. When the London team rejected the Washington wording, Stimson's special assistant, Harvey Bundy, cabled that "after discussion with Groves feel that it would be desirable if the language made clear our position that to a controlling extent this is not considered by the parties a trade or commercial agreement. This position becomes more forceful with the background of two wars and a second anticipated liberation of Belgium. . . ."[27]

If this was not a trade agreement, what was it? Bundy implied it was political, and he was no doubt right. But to insist on its political nature posed the necessity of congressional and parliamentary approval when the war was over. It would give the Belgians an opportunity to postpone any long-term promises and raise the tariff issue again; it would also run the danger of public discussion of secret information and of possible rejection by the Belgian chambers of any previously agreed upon arrangement. The Americans could not have the benefits of both kinds of an agreement.

At a 16 August meeting the Belgians submitted their own counterdraft prepared by De Vleeschauwer. It was shorter and simpler. The section dealing with the first deliveries held no substantial differences from the English version. But for the postwar period, De Vleeschauwer wrote:

Subject to the international obligations of Belgium and the Belgian Congo, considering that the maintenance of world peace could be dangerously affected by the disposition of [uranium] and [thorium] ore, and estimating in consequence that control of these ores should continue after the end of the present war against Germany and Japan, the Three Governments declare themselves ready to reach an understanding with a view to the acquisition and common utilization by the Three Governments for a period of 99 years of those ores produced in the Belgian Congo.[28]

For the allies, the Belgian proposal for common utilization came as a shock and was "obviously unacceptable."[29] At the August gathering Anderson quickly pointed out that his own phrasing assured Belgium participation in any benefits accruing from commercial and industrial exploitation of Congo uranium. Prime Minister Pierlot responded that the Belgian ministers had to be satisfied that any agreements would be in accord with the sentiments of the Belgian chambers and people.

The Belgian ministers' main preoccupation was not with the benefits of their commercial exploitation but with the same question of security to which the United States and the United Kingdom Governments attached such importance. Both for this reason and because Belgium had always occupied an honourable place in the field of scientific research, they were anxious that the industry, workers, technicians and scientists of Belgium should have a part in the work of development. . . . The basis of the agreement would not be one of payment for services rendered, but of complete mutual trust and collaboration in the interest of the common security of the three countries. If the minerals in question were bought at controlled prices and used in common by the Three Governments, this would be achieved and the question of an option or first refusal would no longer be relevant.[30]

Anderson protested that the two powers had expended vast efforts in research, and now the Belgians were, "in effect, taking the line that they would only agree to let the Two Governments have these minerals for military purposes if Belgium were admitted to full participation in the working out of the military problem and its application."[31]

Gutt retorted that "a small country such as Belgium must be particularly careful to preserve its dignity and sovereign rights." The Belgian ministers had approached the matter in friendship, but the British draft seemed based on "unilateral mistrust." He objected to the provision that Belgium could take certain actions, such as use of uranium for the ceramics industries, only if the United States and the United Kingdom approved. The ore might not be worth much without the research of the two powers, but that research "could be of little value if they were denied access to the richest uranium deposits in the world." As for methods of production in the Congo, inspection of the mines, and so forth, "the safeguarding of Belgian sovereignty should be the first, if not the only, consideration."[32] Pierlot, for his part, argued that the distinction between industrial and military uses was unreal. He did not see how Belgian participation would endanger secrecy or security.

The discussion was protracted and warm. Betts suggested that Belgium alone might determine the amount of ores to be used for industrial purposes in Belgium, and Anderson volunteered that normal uses could include anything but use of the ores as a source of energy. Gutt accepted these revisions as helpful but insufficient. Winant commented that British leasing of bases in the Caribbean to the United States— at a time when the Americans were neutral—was a far greater sacrifice than Belgium was being asked to make. He also made clear that the United States could not see the Belgian ores as justification for full Belgian participation, and would do without them. His British counterpart warned that Belgium's position of holding material of vital military importance "might not be an enviable one."[33]

Neither side gave way. Pressure mounted, and Gorell Barnes thought that Pierlot seemed a bit shaken at the end. The Belgian prime minister tactfully offered to search for some compromise which might involve giving the United States a first refusal for military and strategic purposes, reserving for Belgian industry as much ore as required, and promising not to sell any ore elsewhere. De Vleeschauwer also suggested that the contract for 3,440,000 pounds not be held up; perhaps a separate agreement in the form of exchanged letters could be arranged in advance of the main agreement. To this there was common assent.[34]

Had the Belgians, after all points at issue seemed resolved at the end of July, changed their position? Not really. Their earlier draft had mentioned common utilization of the ores by the governments. Moreover, while Gutt and the others had indicated willingness to cooperate, they had never promised an option of first refusal to the two powers. They had only promised to see that the ore did not get into the wrong hands. At the 28 July meeting, when the postwar period was considered, discussion had at first focused on the new issue of the free-trade obligations of the Congo and then on the authority of the Belgians to protect the security of the ores by means of an agreement with the two powers. The Belgian interest in joint participation had not been specifically discussed. The American and British representatives, even in their preliminary conference, had not picked up on the Belgian point; rather, they had been distracted by concern that the Belgian ministers might claim lack of legal authority to make any postwar obligations at all. Anderson and Winant were preoccupied in determining if it would be necessary to employ the prepared "fall-back" wording. The British and American oversight also resulted from an apparent assumption that the initial Belgian reference to "common utilization" of the ores meant their use by the Belgians in the ceramics industry. The representatives of the two governments clearly did not conceive that the Belgians would go so far as to demand participation in the secret military energy researches undertaken for so long, in such secrecy and at great expense, by the Americans and British.

Why did the Belgians not press the issue more strongly themselves on 28 July? Perhaps because Spaak, an experienced foreign minister acutely aware of Belgium's traditional reliance on precise adherence to legal commitments, was himself preoccupied with making his point regarding the free-trade treaties. Perhaps the Belgians thought those treaties might negate for them the necessity of talking about a postwar arrangement, thus avoiding an embarrassing squabble with their friends.

In terms of the general Belgian position, it probably was not significant that Foreign Minister Spaak was not present at the 16 August confrontation. Yet his absence was felt in that Gutt took a larger role as Belgian spokesman. Although De Vleeschauwer drafted the August counterproposal, Gutt took the lead in defending it. De Vleeschauwer scarcely participated in later negotiations, and Spaak was to note that the bonds between himself, Pierlot, and Gutt were stronger than the bonds any of them had with the minister of the colonies.[35]

The sixty-year-old Gutt was an experienced financier and diplomat, physically small and equally determined and feisty. He was jealous of his nation's sovereignty, and this appears to have dominated his thinking. If the British and Americans were asking too much, Gutt went too far in his own phrasing and provoked a reaction that Spaak might have avoided and that Pierlot recognized had to be soothed. To suggest that cooperation might be refused and to accuse allies of mistrust in the very days when these allies were dying in the process of liberating Belgium was not tactful. Nor would Belgium profit were the United States to turn elsewhere for its supplies.

To the British and Americans, Belgian demands for access to military research results after the war seemed an affront. There was basis for their point that significant industrial and military applications could not be separated because they both were founded on energy production. Yet not to accept the great powers' definition of a proper borderline between commercial and military uses was an even greater expression of mistrust than any statements attributed by Gutt to the British and Americans.

Behind the talk of rights of sovereignty and the Belgian resistance to visits to the mines by British and American agents to determine what equipment should be ordered and to monitor the progress of production there lay further issues of mistrust. Sengier had objected to such visits on the ground that the presence of "fat-salaried" Americans would be disruptive. Another objection was never formally mentioned. The price of the ore was to be based on costs of production plus a reasonable profit margin and an allowance for the Canadian radium market issue. Costs of uranium production in Colorado and Canada were far higher than in the Congo, primarily because of differences in wages and because of the difficult accessibility and low grade of the ores in North America. In 1939 most native miners in Katanga earned no more than ten cents per day.[36] Sengier was negotiating for a cost of production that, if not totally based on North American costs, would at least be strongly influenced by them. The Americans thought the price should be determined by the true cost of production in the Congo. American experts at Shinkolobwe could, of course, draw their own conclusions as to what the Belgian expenses were. Sengier's comment also suggests that he was sensitive to the possibility that the Americans might be critical of Union Minière labor policies.

The sole positive result of the 16 August session was the decision to proceed with a separate contract for the first 3,440,000 pounds of ore. The following day, Gorell Barnes drafted letters to be exchanged by the three powers at the time the contract was signed by Sengier for African Metals and by Groves for the Combined Development Trust. The letters were to accompany a straightforward five-article Memorandum of Agreement which stated that the Belgian government approved the price and ensured delivery while the U.S. and U.K. governments ensured acceptance.

Stimson informed Roosevelt a week later that the Belgians "have not yet been willing to make a satisfactory agreement for the postwar period. They are now asking that Belgium be taken in as an equal partner. . . ." Since the Belgians had agreed to a contract for 3.5 million pounds, Stimson advised that Winant be authorized to approve it "if this proves to be necessary in order to get quick action on the development of the Congo supply."[37]

Compromise: Article 9

All seemed in place for the specific tonnage contract to be signed; not until then would the letters be exchanged. But now Sengier dragged his feet. On 2 September 1944 Brussels was liberated by the allies, and the head of the Union Minière began talking about the need to return there, perhaps by way of Paris. That same day the war department telegraphed its negotiators that "every effort" was being made to push through the Sengier contract.

> We have all feared that 51 [Belgium] might be satisfied to rest on this contract, using it as reason or excuse for delaying satisfactory post war arrangement. Since powers of 9 [Great Britain] and 60 [United States] to bring persuasive pressures are perhaps

greater now than later we hope that every possible effort will be made to complete post war understanding at earliest possible moment. . . .[38]

Such efforts were indeed being made. Betts and Barnes met with De Vleeschauwer with little result. Winant called Washington and gained support for his view that negotiations should be completed, getting as much time as possible although the extremely long-term arrangement would have to be abandoned. On 5 September Winant and Anderson spent two and a half hours with Pierlot, Spaak, and Gutt. Winant described the latter two as having "been progressively more uncooperative in coming to terms with us since we refused their proposal of joining us in this matter for common defense purposes which they thought would give them prestige at home." Gutt, on whom the British were counting, in Winant's opinion "rather made matters worse for us than better. . . ."[39]

Initially the three Belgian ministers rejected any notion of first refusal, limited any possible agreement to six years, and claimed as much ore as they desired for whatever purposes, although they would sell no ores to powers other than the United States or United Kingdom. Intending to return to Brussels early the next morning, they probably felt a strong need to be able to demonstrate to compatriots who had suffered in occupied Belgium that while in London the ministers had been good stewards of their national patrimony.

Anderson bore down, and Winant confessed to being "very blunt as their unyielding attitude made it necessary."[40] The American thought the Belgians were a bit shaken by this firmness, realizing that behind the phrases of diplomacy some real frustration and anger were building. They no doubt were also aware how much they needed allied sympathy, support, and supplies for their country. There were other small countries such as the Netherlands looking for allied support. French resentment over the abrupt surrender of the Belgian army in 1940 was still strong, and the French too would be vying for Anglo-American economic support. The Belgian ministers knew they would be returning home and could plead that, after consultation with persons there, further revisions would be necessary. In any case, the Belgians eventually compromised. They accepted a first-refusal arrangement for a period of ten years following completion of the delivery of the first 3,440,000 pounds of uranium oxide. Belgium would be able to reserve "such reasonable quantities of the said ores as may be required for her own scientific research and for her own industrial purposes exclusive of any process involving the use of such ores as a source of energy except as specifically provided."[41]

The allowance of ore for scientific research was a new and reasonable concession which respected the pride of the Belgian scientific community while protecting the allies' concern for energy research. Yet Belgian commercial use of the ores was not forever to exclude energy. The special provision as phrased in what became Article 9 of the final agreement read as follows:

(a) In the event of the Governments of the United States of America and of the United Kingdom deciding to utilize as a source of energy for commercial purpose ores obtained under this agreement the said Governments will admit the Belgian Government to participation in such utilization on equitable terms.

(b) The Belgian Government undertake that, in the event of their contemplating the use of such ores as a source of energy, they will so use them only after consultation and in agreement with the Governments of the United States of America and of the United Kingdom.[42]

Winant and Anderson were convinced that this formula was the best obtainable from the Belgians, and the war department later concurred. Only that night did Winant discover that the Belgians planned to leave early the next morning on a British military flight. He quickly informed Anthony Eden, the British foreign minister, who arranged for the flight to be delayed so that one last gathering could be held at the foreign office.[43]

Eden and the Belgian ambassador, Baron de Cartier, were present for the first time. A Memorandum of Agreement drafted by Barnes and Betts covering both the specific tonnage agreement and the postwar years was discussed at length. No major problems arose, yet the Belgians were reluctant to sign despite the presence of Eden, which must have been intended as further pressure for conclusion of the accord. Finally it was agreed that one of the ministers would return from Brussels to London authorized to sign a formal agreement. Meanwhile Spaak, Eden, and Winant would initial the draft to make the understanding a matter of record. Pierlot, in explaining his reasons for not signing, said he and his colleagues needed to garner more information about thorium and wanted a clear definition of Article 9(a). Eden, to assure that other issues would not later arise, had these reservations typed and read to the ministers for confirmation. Winant observed:

> I felt we had pinned the Belgian's [*sic*] position by having their Minister of Foreign Affairs initial the memorandum of agreement or I would have objected to those verbal reservations and demanded signature to a completed document. . . . We would have had to begin negotiations again at a later date of the after war phase I am sure if the Belgians had returned to their country without formalizing our understanding and perhaps the Belgian Government's guarantee of the contract [for 3,440,000 lbs.] would have been jeopardized.[44]

With this hurdle cleared, the next step was the signing in the United States of the contract with African Metals. Still Sengier delayed. When informed on 19 September of Sengier's desire to return from New York to Brussels, Winant and Anderson advised their colleagues in America to facilitate the trip, but they opposed Sengier's request to visit Paris. The two pressed for Sengier's prompt signature. Later that day the diplomats reversed their position. If Sengier insisted on meeting with his government and company officials, they suggested all parties should be brought to London and an attempt made to conclude the intergovernmental arrangement. Even this involved the risk that Sengier might persuade what by then would be a new Belgian government to go back on the whole deal. Utmost efforts should be made to get Sengier to sign in the United States, and he should not be told that he could have a visa to Brussels.[45]

That various forms of heavy pressure were being contemplated in Washington is evidenced by a memorandum Bundy drafted on 18 September to send to Secretary of War Stimson. In it he pointed out that negotiations with Belgium had gone well until the country was liberated. Now all that could be obtained was a ten-year option, and the Belgians were still delaying and expressing reservations. The Belgian government, however, was currently attempting to negotiate large reconstruction loans with New York banks. Similar Dutch loans had been arranged with the "approval and blessing" of the state and treasury departments. Bundy wrote:

> It would seem wise for this Government to delay approval of a large Belgian loan until we know:
> a. Whether the Belgian Government will sign the present agreed option.

b. Whether the granting of such a loan should be made condition-
    al upon an extension of the option for a longer period of time.[46]

The memo was canceled; for what reason is unclear. Perhaps Bundy
realized that such blunt pressure might make the proud country even
more resistant. Perhaps it was because negotiations with Sengier were
finally making progress.

The African Metals Corporation Contract

Some of the issues which had slowed negotiations on the contract with
the African Metals Corporation were points that had been raised in the
diplomatic discussions, such as the presence of American observers at
the mines and the marketing of radium. Others issues were more general, reflecting not so much disagreement as different approaches.
Groves' tendency to push for fast fulfillment and to nail everything
down in writing somewhat put off Sengier, who believed that the success of their previous dealings and the concept of good faith should
govern.

Sengier had met with Sir Charles Hambro of the British Raw Material Division and presumably also Groves in New York on 5 July. The
next day he sent them a written version of the basis of accord reached
at that meeting. It was not until the end of August, following the series
of meetings in London, that the Belgians received a detailed contract
from the British and Americans. By the time Sengier made his response of 5 September, Brussels had been liberated, and the businessman had second thoughts. He had made concessions in early July, he
wrote, because of the urgency with which the purchasing powers had
called for rapid agreement for military reasons. But the long delay—
not attributable to African Metals, he noted—caused him to wonder
whether the concessions should have been made.[47]

Sengier saw the proposed contract as significantly different from
what he had accepted in July. He felt he had already made concessions
then, agreeing to reopen the mine on the basis suggested by the Americans in February, despite the objections of his managers in Africa and
not withstanding the inconveniences involved. Moreover, there was no
news regarding the general tripartite accord, to which he had given assent, provided that four matters of principle were taken into account:
(1) the difficulty of meeting extraction goals because of the lack of min-

ers, (2) the right to reserve production for the ceramics industry and other industrial applications, (3) a promise from the purchasers not to penalize the rights of entry for radium and other by-products of the uranium mining, and (4) no interference in prospecting or exploitation, no permanent observers in Africa, and no foreign mine workers "who will provoke social troubles."[48]

In particular, Sengier found inadequate the proposed price, the price and percentage of restitution guaranteed for radium, the method of calculating the amount of radium per ton of oxide, the amount and method of indemnification for the expenses of reopening and pumping the mine, the absence of an indemnity should the contract be forfeited before all the oxide was delivered, the means for determining the value of minerals lost in transit, and the insistence on various controls and measurements at the minehead, which he saw as interference in the operation. Only if these points were met could he feel able to sign without returning to Brussels to consult his colleagues. In a postscript Sengier suggested that the newly mined ores be refined at Belgian plants rather than in the United States, thus simplifying the whole matter of returning the leftover sludges.

Groves quickly had Sengier's letter reviewed. The report received on 8 September argued that the UM executive's points regarding the ceramics industry and radium had been dealt with in London and in any case could not be part of the contract under study. As for mine labor, the problem had been taken into account; the Belgian draft by Mr. Leroy had been altered on this point only so that inability to obtain miners would not give UM an excuse completely to avoid the contract. As for an observer, it was customary for buyers to have a representative on the scene, and the complicated nature of the contract required it; he would be an army officer instructed not to provoke social troubles. The oxide prices corresponded to the Belgian draft: $1.45 per pound for material containing 50 percent or more oxide, $1.40 per pound for material containing less than 25 percent oxide. No price was fixed for the materials in the 25 percent to 50 percent range; a compromise figure of $1.425 per pound was suggested.

The radium issue was a technical one involving measurements of residue; no yielding on price was recommended, but some explanations and possible solutions regarding measurement were put forward. As indemnity for costs, the Belgians had first suggested $350,000, but then revised their estimate to $400,000, payable immediately upon signing. The Americans held to the first figure and argued fairness in that this amount should be paid in steps as the work was completed. The 31 cents per pound indemnity for ore undelivered should the contract be forfeit, though in Leroy's draft, had not been discussed in July and therefore omitted; the London accord, however, would guarantee the contract. Values of lost minerals were covered by the proposed contract if interpreted correctly. Groves' office rejected the implication of any intention of interfering with the UM operation in Africa and strongly opposed consideration of any refining in Belgium at that time—perhaps later.[49]

The stage was set for another bargaining session in New York on 11 September. At this meeting Sengier was apparently convinced that most of his concerns were being met either at London or in the contract. He may have remained uneasy about the observer, but his concern was no doubt assuaged by a concession that the $1.45 price per pound would hold for all materials with oxide concentrates above 25 percent rather than 50 percent. This was a greater compromise than initially envisioned by the Americans and in the long run would be worth many thousands of dollars. Some of the step schedules for payment of the $350,000 indemnity for mine repairs and dewatering were reduced, as were the steps for the $200,000 indemnity for electricity diverted from other Union Minière projects. A few more wording changes were later accepted, and the seminal contract was complete.[50]

Despite the apparent agreement on detail, Sengier still delayed, and a week after the last New York meeting he made his request to return to Brussels. This move no doubt was prompted by the need to coordinate with whatever the government was doing. If the contract were to be guaranteed by a tripartite accord and Sengier essentially covered by the government's approval of that accord, he needed to know whether or not that accord was going to be signed. He had no doubt learned of the agreement hammered out and initialed in London; this knowledge had made it easier for him to ease on 11 September some of the reservations expressed in his letter of a few days earlier. But final commitment was another matter. It is reasonable to assume that Sengier received some sort of reassurance in the days before his signing; perhaps word that Sengier had worked out an acceptable contract with the Combined Development Trust similarly encouraged Spaak and Gutt. Sengier signed the contract for sale of 3,440,000 pounds of ore on the 25th. On the 26th, at its final cabinet meeting, the Belgian wartime government authorized Spaak to conclude without alteration the Tripartite Agreement initialled on 6 September.

Sengier had a good contract, though he later would be second-guessed regarding the prices he demanded. When these eventually became known in some circles, occasional critics would say he did not charge enough for his material while it was still rare; others would say he exploited the West. The presence of both charges suggests that Sengier struck a fair balance. Few private figures have had such authority in such an advantageous position and used it with such discretion. At heart Sengier was loyal to the Western democratic cause; at the same time, he acted responsibly on behalf of his stockholders. The urgency with which the purchasers pursued his wares was great, evidenced in the contract by the frequent phrases "as soon as possible," "diligently prosecute," and the like. If Sengier found the wording of the first paragraph of Article 1 not to correspond to the way things were done in Africa, the insistent phrasing reflected the leverage the British and Americans knew Sengier had:

> . . . the Seller hereby agrees that it will make all reasonable efforts and cause to be made all reasonable efforts to procure sufficient personnel, material and equipment and do all other things necessary to produce diligently and expeditiously and deliver promptly to the Buyer the Q-11 contained in ore . . . which the Buyer herein contracts to buy.[51]

With the signed contract and the initialled accord in hand, the Anglo-American team felt a sense of accomplishment, but final assurance that all was settled did not come quickly. Not until 11 October was word received from De Vleeschauwer that the previous cabinet had approved the accord. Because it was thought not desirable that the question be brought before the new cabinet, the exchanges would be dated 26 September.[52] All four of the ministers who had been instrumental in the negotiations were continuing in key posts in the new national union government headed by Pierlot. As it contained Communist party representatives, however, serious debates might have arisen and therefore the path of attributing definitive action to the previous cabinet was chosen.

Conclusion

The negotiations had been arduous and took far longer than originally anticipated. The reasons for the haste of the British and Americans varied over the years. In 1942 and early 1943 they were chiefly concerned with securing enough uranium so that research could be carried on. As the possibility of a bomb grew and reports were received of Nazi atomic research, the need to preempt the market became clear. If the Belgians were unaware of the value of their ore, it was possible that some might reach Germany. But these could no longer have been the chief considerations by the summer of 1944, when pressure for a quick settlement began to build.

By that time the Americans had enough ore for the expected duration of the war, and they knew Belgian ore was not reaching Germany. There was general fear of what might happen if atomic capabilities proliferated in the chaos following the war's end. No specific comments regarding the Soviet Union appear in the written accounts of the negotiations thus far available. Yet if the diplomats were concerned only about Germany and Japan, which enemies they had no difficulty in naming, why the frequent references to "wrong hands" unless the phrase was intended to veil indication of a current ally? If Groves' comments of years later are taken at face value, he apparently believed very early that his chief opponent was Stalin, not Hitler: "there was never from about 2 weeks from the time I took charge . . . any illusion on my part but that Russia was our enemy and that the project was conducted on that basis."[53]

The Americans and British may not yet have been greatly worried regarding the progress of Soviet atomic research; still, the tensions which would produce the Cold War were growing. The effort to monopolize uranium supplies was a natural protective step which no doubt was not ignored in Moscow as the flow of ore to the United Kingdom and the United States became evident.

Groves opposed the Russians and admitted a sense of competition with the British. For the French, he had distrust, especially regarding how they might allow information to reach the Soviets. Joliot-Curie's approach to Sengier before the war reflected his awareness of the possibility of atomic fission. As far as the Americans were concerned, Joliot-Curie was both a potential competitor and, as a Communist sympathizer, a security risk. In later years the United States would strongly resist the Frenchman's efforts to become associated with the Belgians in uranium-related research.

Both sides used timing to their advantage in the negotiations. The Belgians, like many small powers pressured by greater ones, maneuvered for time. They saw that their friends' immediate needs were met, were concerned about how the Belgian public might view their actions later, and perhaps thought their position might grow stronger as Belgium was liberated and the end of the war approached. In fact, the importance of continued allied friendliness became greater during the days of liberation, for the divisions that wracked Belgium emerged more fully as the common enemy withdrew and economic chaos prevailed. Winant, Bundy, Eden, and Anderson saw this and sensed the psychological and political need of the Belgian ministers to hasten to their country to deal with even more pressing issues. By their insistence the British and Americans pushed to conclusion negotiations that might have dragged interminably.

The Belgians acquitted themselves well, obtaining a sure market for their uranium at a price the diplomats suspected was high. They won compensation for loss of radium ores, as the United States returned the sludge to African Metals after its processing for uranium oxide. The Americans agreed to equip the Shinkolobwe mine, and the contract term was limited to only ten years following the delivery of the first 3,440,000 pounds (i.e., to c. 15 Feb. 1956). Thus Belgium gained far more than the great powers thought they would have to concede when negotiations began. This ability to resist demands of great powers is often considered one of the best measures of the diplomatic success of a small state.[54]

The United States and United Kingdom fared well also, in that they did get control of the Congo uranium output, if only for a little more than ten years. They soon estimated that rapid excavation might exhaust the Congo reserves within that time period anyway. The combined prices also successfully avoided the radium tariff issue. Concessions on the energy issue were greater than desired but kept the pace of revelation of such research in their own hands. The stubbornness with which both sides negotiated this issue helps to explain why it remained a sticking point in later years.

Pride and testy defense of its sovereign rights are hallmarks of the small nation's diplomacy over the years.[55] Gutt's obstinacy did protect Belgium's rights; yet in thinking to take the ball home and not let the others play if the game were not played his way, Gutt let his pride carry him too far. His words served the Anglo-Saxons in that they revealed how serious the Belgian resistance was and at the same time relieved the allied negotiators of feeling bound by certain restraints of gentility in dealing with a defeated and sensitive small power. Anderson and Winant apparently never threatened any faltering of support for Belgium's needs in the coming months. They simply turned the Belgians' traditional argument of loyalty and dedication to a common cause to their own use. In the long run, Pierlot, Spaak, and even Gutt had to respond to this. Moreover, the Belgians basically did wish to cooperate over the long range. They had a ready excuse in the form of lack of authority as a government-in-exile to commit the country after it was liberated; though Anderson feared they would take this route, they did not. After all, Belgium had no desire to deal with the Russians or Germans, and any possible French interest was remote.

Both sides initially wanted too much. The American and British concept of perpetual control of the Congo uranium was contradictory to the spirit, if not the letter, of Belgian sovereignty and conflicted with international agreements regarding free trade in the Congo Basin. The limit of 99 years scarcely improved the matter; the compromise span of ten years did. It was surely within the ministers' authority and their nation's sovereign right to dispose of their ores for a reasonable but limited period of time. The qualifying clause "subject to . . ." technically cleared the conflict with the free-trade treaties, though purists could argue that the conflict remained. An important shift had occurred, however, in the adoption of the final wording. The Belgians were now controlling the sale of the product, rather than agreeing to surrender control to others.

It was going far—and dangerous, as Betts warned—for the Belgians to demand equal participation in the British and American research after the war and thus have access to much of what was discovered during the war. The importance of the need for secrecy and uranium control became more clear to the Belgians after Hiroshima and the difficulties which arose with the Soviets following the armistice. Nor were the ministers aware of the restrictions on exchange of atomic information even between the Americans and the British. Thus they did not know the extent of what they were asking or how it would require revision of British-American arrangements. These arrangements, after all, called for interchange only in areas where both countries were working. Belgium had no atomic energy research program, yet Gutt thought she should have access to Anglo-American research results. It was unlikely that Anderson, who had labored hard in framing the Quebec Agreement, or Winant could or would have given way on this.

The combined Anglo-American approach increased the Belgians' desire to cooperate and weakened their will to ask a higher price. To their credit, the ministers' price was not one of financial reward but rather of respect for their country's dignity and commercial future. They did support Sengier's concern regarding radium tariffs but accepted compromise and did not get into the price-per-pound negotiations. In its final form the Tripartite Agreement did not mention price at all. This was left to Sengier and Groves; although Sengier preferred a market price proposal, he had to negotiate the matter without ministerial backing, and the cost-plus solution prevailed. Groves, who with Hambro negotiated for the Combined Development Trust, was more willing to part with cash than were the diplomats, so Sengier did achieve a good price. Groves' willingness to pay reflected his view of the necessity of gaining the ores and perhaps also the tendency, often bewailed by civilians, of the Corps of Engineers not to be so cost conscious as private entrepreneurs.

For the Americans, British collaboration was invaluable. It came about more for negative than positive reasons, that is, a desire to prevent a British monopoly of uranium produced in the Congo. Yet in the end it demonstrated more the strength of the British-American tie than weaknesses within it. Britain gained too, for it appeared she was obtaining access to all the uranium she needed and could afford, while being protected from its falling into the hands of powers far less friendly than the United States.

A crucial yet shadowy figure influencing all the dealings is Edgar Sengier. Able and far-sighted, he was relied upon by both the Belgian officials and the American military. Groves considered him a key colleague whose attitude and cooperation were of the highest order and of inestimable value. Surely the West owed him great thanks for his foresight in assuring the safety of existing uranium supplies in 1939, for calling them to the attention of the Americans, and for arranging their sale so quickly and without fuss. Yet if he supported full collaboration in the war years, he was a strong spokesman for a free hand in the postwar years and a key delayer. Groves was slow to realize this, if he indeed ever did. On the other hand, diplomats in London and Washington came to see Sengier as a possible sticking point more difficult than the Belgian ministers who might be influenced by broader considerations than the Union Minière executive. Had Sengier been willing to sign all the contracts proffered without referring the matter to his government, the diplomatic negotiations might never have occurred. His alertness in this matter was of high service to Belgium. Sengier did know well how to work with the Americans, and his intermediary role in subsequent years would prove salutary to all three nations.

The Tripartite Agreement of 1944 was the cornerstone of the Anglo-American source procurement program. In time other sources would be found and agreements negotiated, but this was the crucial one. The ores which Sengier brought to New York before the outbreak of war were of inestimable value for the American research program. Without them, it is doubtful that atomic devices could have been prepared by 1945; the course of the war in the Pacific and many subsequent events would have been far different. The tons supplied in the initial postwar years were to fuel the development of the American nuclear shield which in turn was to have extensive meaning for the balance of power in the new Europe. As one Western diplomat put it,

> The German invasion caught them (the Russians) still unready and swept them to what looked like the brink of defeat. Then came the turn of the tide and with it first the hope and then a growing belief that the immense benison of national security was at last within their reach. . . . There was great exaltation. Russia could be made safe at last. She could put her house in order and more than this from behind her matchless three hundred divisions she could stretch out her hand and take most of what she needed and perhaps more. It was an exquisite moment. . . . Then plump came the Atomic Bomb. At a blow the balance which had now seemed set and steady was rudely shaken. Russia was balked by the West when everything seemed to be within her grasp.[56]

Such a version may be overdrawn and overdramatic. Yet the importance of continuing Western nuclear research and weapon development assured by the Belgian agreement cannot be underestimated.

What was achieved by the prolonged negotiations was a cornerstone also because the final document, though carefully kept secret for a number of years, was inevitably to guide and influence other negotiations. Lessons learned in the discussions with the Belgians would be reflected in the sorts of approaches taken toward other nations by the British and Americans. Because the Congo ores were so central to their program, the Americans would always have to take care that other arrangements, no matter how desirable, did not bring concessions that had not been granted to Belgium. Where the talks with Gutt and Spaak had proved difficult lay shoals that would endanger other proposals; failure to take these into account would in time cause further problems.

3 · *Efforts at Preemption: Brazil, The Netherlands, and Sweden*

Assured access to Belgian ore greatly eased the position of General Groves and the Combined Development Trust, but it did not cause them to relax their efforts. The success of the Belgian negotiations, the size of the contract, and the depth of the cooperation achieved instead encouraged them to contemplate more sweeping arrangements. The goal now became twofold: to fuel the research program being conducted in North America and to gain world-wide control of all substantial supplies of uranium as a means of deterring development of the bomb by others. Uranium was thought to be a rare entity and even more rare in the sort of concentrations that existing production facilities could utilize. No one knew exactly what deposits existed in Russia; for the most part they were not considered to be good. The Czech mine at Joachimstal would supply the Russians with some ore, but perhaps not enough.

The Survey of Uranium and Thorium Resources

Initially the concern was that Germany rather than Russia might gain access to the rare ore. As a result of the MAUD Committee report, the British ministry of aircraft production prevailed upon the ministry of economic warfare to take steps to prevent such an occurrence. The source most accessible to the Germans outside the region of their own domination was the low-grade reserves in several locations in Portugal. A sum of 500,000 pounds sterling was made available to the British ministry, and it promptly bought options on a number of Portuguese concessions through its commercial arm, the United Kingdom Commercial Corporation. The options permitted more extensive geologic explorations. These in turn indicated that most of the Portuguese mines were of little value, and the options were dropped. A contract was signed with one Portuguese firm for all the uranium and radium it would produce for two years beginning in March 1942.

The most important Portuguese site was the Urgeirica mine. Negotiations were difficult and protracted; eventually a contract was signed in May 1943. Even in this case, the yield was not expected to be substantial. In London, the total Portuguese output of uranium per year from all sources was estimated to be no greater than forty tons. Some shipments were made to England, but the supply was so negligible compared to the amount needed for meaningful research that there was little objection to the few active Portuguese uranium mines finally being placed on a care and maintenance basis. After all, the British could obtain all the ore they needed from Canada, or so thought the British department of scientific and industrial research, which in early 1942 surveyed world uranium resources.[1]

Groves himself had initiated some minor efforts to determine sources of uranium in 1942. By early 1943 it was clear that a major study was needed. Extensive investigation by a government agency, especially in the field, would arouse suspicion. Groves therefore wrote a contract in May 1943 with a subsidiary of the Union Carbide and Carbon Corporation, a firm already connected with the Manhattan Project as a supplier of graphite and as a designated future operator of a gas diffusion plant. Working closely with this subsidiary, the Union Mines Development Corporation (UMDC), was the Murray Hill Area, which was the Army Corps of Engineers office concerned with uranium discovery established in June 1943. Major Paul L. Guarin, a former geologist with the Shell Company, was tapped by Groves as area engineer and army liaison to the Union Mines Development Corporation. Among his key assistants were Dr. George W. Bain and Dr. George Selfridge, geologists from Amherst College and the University of Utah.

Groves made clear he wanted a sound report but did not want to wait forever for a 100 percent perfect one. An extensive effort was promptly launched by a staff of geologists and linguists specially selected by the UMDC. Bibliographic research covered some 67,000 volumes, over half in foreign languages, resulting in fifty-six geological reports concerning over fifty countries. Field explorations were made in thirty-seven countries by geologists carrying no identification linking them with the U.S. Army or the Manhattan Project. Property was acquired on the Colorado Plateau for the U.S. government and in the Great Bear Lake area of Canada by a U.S. Army backed firm called Ventures, Ltd., on behalf of the UMDC. Boris Pregel was also active, staking claims for his International Uranium Mining Company which he subsequently tried to peddle to Groves. The Canadians were not happy about these explorations and Groves' office soon admitted that the United States should not have been surveying in Canada. The Ventures, Ltd., claims were eventually assigned to the Canadian government following its decree of 15 September 1943, which reserved to the Crown any radioactive substances henceforth produced in the Northwest Territories and the Canadian Yukon. On 28 January 1944, Eldorado Mining and Refining Ltd. finally became a Crown company.[2]

The United States and the United Kingdom were duplicating each other's efforts and working at cross-purposes in the tracking down of available uranium supplies. The Quebec Agreement between President Roosevelt and Prime Minister Churchill and the subsequent creation of the Combined Policy Committee (CPC) and the Combined Development Trust (CDT) inaugurated, however, a period of better cooperation on atomic affairs. On 24 August 1944, Sir Ronald Campbell proposed to the joint secretaries of the CPC a coordinated program for procurement of raw materials, and in particular a "comprehensive and detailed fact-finding study of supply and demand aspects of raw materials" to be undertaken as a matter of urgency.[3] This seemed all the more appropriate as allocation of these materials between the allies was a responsibility of the CPC. Apparently the British at this time were not fully informed of the UMDC project or, more likely, were aware of it and, sensitive to their own lack of manpower to undertake a duplicate massive effort, desired joint participation. So it was that the Committee was formally charged on 19 September with the responsibility of making a survey of world sources of uranium and thorium.

In practice this meant that the ongoing research efforts of the UMDC and Murray Hill Area were augmented in the fall of 1944 by three British experts. Despite their late arrival and few numbers compared with an American team ten times as large, the skills of the British won them acceptance, and the report produced in October 1944 would be greatly shaped by British drafting.

The report, significant in itself, was not the only achievement of the UMDC effort. Another was the development of an easily portable and practical Geiger-Muller counter which vastly accelerated prospecting and in time would play a key role in making the report obsolete. Two theoretical developments by Professor Bain were even more important. From pattern studies of conditions under which uranium was found, the Amherst geologist determined that the ore might be associated with hydrocarbon-bearing materials (coal and oil shales). This greatly expanded the areas of search.

Bain also argued that uranium might be found in monazite sands. Research on the location of these had only just modestly begun as a result of recent scientific developments suggesting the utility of thorium, an element most frequently found in monazite, for augmenting production of fissionable uranium.[4]

The UMDC survey was submitted by the CDT to the Combined Policy Committee and received the latter's approval on 26 October 1944. Section 1 of the report contained a fourteen-point summary of the conclusions and recommendations of the CDT and its survey team, all of which were accepted by the Committee. The Congo's dominant position in uranium supply was emphasized, along with the desirability of a close understanding with the Belgian government, the Union Minière, and the chief official of that company. The next best source of supply was seen as Canada; after appropriate development over ten years it might produce a quarter of the world's output. American deposits were described as low-grade, costly to mine, and not worth developing if the assets of the Congo could be obtained. The only other deposits of significance then known were in Russia, Czechoslovakia, and on a lesser scale in Australia, Portugal, Madagascar, and Bulgaria. Though Russia might have further developed her radium mines—thus increasing access to uranium—it was estimated that the Russian deposits joined with those of Czechoslovakia and the rest of Eastern Europe would constitute no more than 5 percent of the world supply of uranium once Canadian and Congo production was in full swing.[5]

The potential of Australia required further exploration; the surveyors thought it to be less than that of Shinkolobwe or Eldorado. The resources of Portugal and Madagascar, though minor, could be dangerous in hostile hands, and therefore preemptive control should be considered. The aid of existing geologic survey teams in British East African territories and India should be enlisted. The UMDC was currently pursuing field investigations in Madagascar, Portugal, Spain, New Zealand, Peru, Bolivia, and elsewhere. Sweden presented a special problem which required investigation at an early date. Similar steps should be taken in Germany as military occupation permitted. More attention also needed to be given to Japan, Turkey, and to whatever could be found out about Russian deposits.

The report next reviewed sources of thorium, listing India and Brazil as the best loci of supply for military requirements. The report recommended that action on these "be sought immediately through political channels" even though the material was not currently required.[6]

Success would give the allied governments a three-to-one predominance, a predominance which could be further strengthened by deals with the Dutch East Indies and Madagascar. Other minor sources of thorium were so widely distributed as to elude control, short of a general international agreement.

Thus development of Congolese, Canadian, and American sources of uranium was seen as giving the CPC control of over 90 percent of the world's "likely supply" of that ore. Similar domination of thorium supplies was impossible, yet control of a major portion of these was conceivable. Therefore

> it will always be possible for a hostile Power to accumulate significant amounts of both uranium and thorium either by extensive buying of small quantities from scattered sources or in some cases by developing resources within its own boundaries. Although the absolute quantities might be dangerous, the proportion of such accumulations to the total world production would be small (perhaps only 5 percent) in the case of uranium, but somewhat larger (assuming Indian and Brazilian supplies were in our hands) in the case of thorium.[7]

The need for continuing assessment of expected requirements was outlined, for apart from the Congo, Canada, and a "slight possibility" in Sweden, the governments were unlikely to obtain additional ores. The production of the Congo, Canada, and the United States was anticipated as being about 1,500 tons of oxide a year. Further exploration and efforts might be able to raise production after five years to about 3,000 or 3,500 tons, but that was a matter of conjecture. The writers of the report concluded with a warning:

> We hope that the strength of our present raw material strategic position . . . will not lull our Governments into a false sense of security. Quantities which may be significant for military production can be obtained by other Powers and the lead which our Governments now have could be appreciably reduced if some other Power exerted itself whilst our own Governments rested on their oars.[8]

A final admonishment exhorted the Combined Policy Committee to move at the earliest possible moment on the several action recommendations presented.

Positive steps were needed in more than just a preemptive sense, if other reports are also taken into account. At the end of December 1943 Lieutenant Colonel John R. Ruhoff of the U.S. Army Corps of Engineers reported that 6,600 tons of U_3O_8 would be required by the Manhattan Project up to May 1946. It was expected that by June 1944 1,426 tons refined from Congo ores and 674 tons from all other locations would be in hand; another 2,327 tons from Belgian ore would be under contract as would 1,694 tons from elsewhere. That left a shortfall of 479 tons.[9] Though the British were taken aback by what they considered profligate American use of the rare material, even conservation might not help because of another problem: shortfall in anticipated production at Eldorado. In order to meet immediate contract demands in 1943 the owners had concentrated on mining and milling the richest and most easily available ores without doing development work in other areas of the mine. By spring 1944 one of Groves' most trusted geologists, Major Phillip L. Merritt, was reporting that production of U_3O_8 was some 27 tons behind schedule as of April and likely to fall increasingly behind schedule as the year progressed.[10]

These problems notwithstanding, Groves estimated that sufficient supplies were in hand, under contract, or expected to be under contract to meet needs until January 1946. He did not fail, however, as chairman of the CDT forcefully to recommend to U.S. Secretary of War Henry L. Stimson, the chairman of the CPC, steps that the UMDC reporters saw desirable for immediate action. Permanent survey groups like that in the United States should be established in England and Canada, with investigations in the areas controlled by the Combined Development Trust divided among them. The resources of the Congo should be exploited by additional contracts under the terms of the new treaty, "conserving as far as possible the resources of North America."[11] The Trust should obtain a purchase option on the Urgeirica, the small Portuguese mine, from the United Kingdom Commercial Corporation, pending appraisal of its value. Similar action should be taken elsewhere when advisable. World thorium resources should continue to be investigated and initiatives begun for "securing control by governmental agreement, if possible, over the resources of Brazil."[12] The reporters did not believe any attempt should be made actually to purchase supplies until more was known about the true value of thorium.

The Trust also adopted a series of assumptions, growing out of the survey studies, to guide its future actions. Given the wide distribution of uranium and thorium, often in relatively minor deposits, it was "impracticable" to obtain control of 100 percent of the world's supplies. A fair goal to aim at would be to control 100 percent of the known major deposits (those of the Congo, Canada, and the United States already constituting 90 percent of the world's total), and 75 percent of the minor deposits such as those in Portugal, exclusive of those in the U.S.S.R. The goal for thorium would be control of 90 percent of the known major deposits (which in turn were estimated to account for 80 percent of the world total) and as many minor deposits as could be secured without difficulty. No efforts should be made to control trace deposits of uranium, and the necessity of uneconomical bargaining in some instances might have to be accepted.

Groves, speaking for the Trust but no doubt reflecting his own views which he tended aggressively to meld with the deliberations of that group, went on to state that "the stock pile of uranium within our territories should be as large as we can make it."[13] Congo deposits should be purchased and moved to safe storage within the territorial limits of the allied governments. Those in Czechoslovakia, Portugal, and Madagascar should be controlled by purchase or political agreement if field investigations so justified; the material should also be moved if economically possible. Minor uranium deposits in Brazil should be controlled by political agreement without purchase or removal; Brazilian thorium should be bought and moved. Thorium from the Dutch East Indies should be controlled by purchase or political agreement and moved if this could be done economically. Control of small thorium deposits throughout the world should be achieved by political agreement without actual purchase or removal.

> While it would be undesirable for any other government to have any stocks of the materials, it would be positively dangerous to permit any other government to secure in excess of one thousand tons of uranium oxide. It would again be dangerous to permit in excess of one thousand tons of thorium plus one hundred tons of uranium oxide to fall under the control of other governments.[14]

These, then, were the goals of Groves and the Combined Development Trust: procurement and preemption. Though the initial concern in the establishment of the survey was procurement of supplies to fuel the nascent atomic research and production program, the tone of the report and its recommendations shows that by the close of 1944 the emphasis was on preemptive control. The aim was to have under the jurisdiction of the Trust 97.5 percent of the world's supply of uranium and over 90 percent of the supply of thorium (the possibility of reserves in Japan and Russia unknown to the researchers was acknowledged). The Americans' program was expanding so rapidly, however, that within a year the prime concern would again be procurement of supplies. Nevertheless, for the first part of 1945 preemption still took first place.

Another change is also discernable in comparing the positions held by Groves and the diplomats in 1943 and 1944. When initial approaches were made to the Belgians, the apparent concern was that the German enemy might obtain the uranium for creation of an ultimate weapon. But by November of 1944, as the allies were sweeping across France, the preemptive concern was no longer directed against Germany—still the current enemy—but against the Soviet Union. The limits of uranium and thorium that the CDT was prepared to let slip through its grasp clearly implied that no other powers should be permitted to obtain the bomb. The military hegemony of the United States and Britain would thus be assured. Such an attitude, were it known by the Russians, could scarcely have encouraged them to believe in the sincerity of Western promises of a desire to work out an equitable division of influence and mutual understanding in the postwar years. Could the Communists expect to be treated fairly when the capitalist states held the supreme weapon and the Communists knew the American and British military were doing all they could to prevent Russia from access to the vitally necessary fuel?

Routine observation of diplomatic activity and ore shipments would in time have demonstrated what the Western allies were up to, but the Russians had more direct and probably prompt information. The man who was British embassy first secretary since 1944 and who would sit on the Combined Policy Committee from 1947 to 1948 was the spy Donald Maclean. He surely had access to the supply reports of the CDT in 1947 and may have seen those even in 1944. Thus the Cold War already developing over a number of issues received significant impetus in 1944 from the search for rare ores, although the full extent of this possibility was not recognized by the West until after Maclean's defection.

Negotiations with Brazil

The second week of December 1944 Groves urged Stimson to start negotiations with Brazil, adding that the department of state would have to be brought in. Groves and the Manhattan District did not have appropriate connections there or any way to establish them. The general's wish to work only through a personal acquaintance or the rumored new ambassador scheduled for Brazil, Adolphe A. Berle, rather than the existing ambassador, Jefferson Caffery, may have provoked the ensuing delay; yet it is likely that British questions were the prime cause of the wait. London learned of the intended approach by accident. Officials there resented the manner in which Groves privately charged ahead just at a time when the United States was reproaching the English for not disclosing some of their arrangements with French scientists. Moreover, the British feared security leaks in Brazil, where there were a significant number of German immigrants and German influence. The British had also been dragging their feet on the need for obtaining control of thorium and had only recently and reluctantly been won over by new scientific developments. An American apology for impetuousness was accepted; the British, aware the United States would move with or without them, came along.[15]

On 6 February 1945 Groves fired a memorandum to Harvey H. Bundy, Stimson's assistant, that was typical of the general's forceful manner. Thorium, Groves suggested, "will be the ultimate means of producing what we are after." Uranium was scarce. The extent of supplies available in the Congo, Canada, or the United States was not known. Thorium was much more plentiful, and its "vital importance" well recognized by all nuclear physicists, including the Germans, Russians, and probably the French. The Germans had tried to take all thorium supplies out of France. "The Russian scientists are, in my opinion, particularly, due to their capabilities and interest in pure science as well as their known espionage of our work, to be [sic] fully aware of the value of thorium."[16] If the Trust could control the supply in India and Brazil, it could "dominate the situation."

> We must not delay. There is no time to lose. We, the British Empire and ourselves, can get control of the Indian supply at any time. Sir Charles Hambro has informed me that this can be done as soon as the Brazilian situation is in hand. No one can be certain of the continued stability of the Brazilian Government. Our President has a strong influence on the present head of the Brazilian Government. We should not lose this opportunity when our Secretary of State will be present in Brazil. I have never understood that it would be other than unwise to have this problem handled

through our present Ambassador to Brazil. I can see nothing to be gained by delay and on the other hand I can foresee grave danger of complete disaster by temporizing and delay.[17]

The security risk involved in approaching Brazil would be no more than that run in talking with the Belgians, Groves thought. After hostilities were over, the risk would be greater because it would be more difficult to negotiate secretly and press for prompt decision. "If there was ever a time for secret diplomatic agreement secretly arrived at, this is the time."[18] It would be disastrous if the same supplies came under the control of Russia or Germany or were simply disposed of on the highest-bidder basis. The general seemed insensitive to British dislike of his determined rushing ahead and attributed London's concern to British lack of representation at any initial conversation. Yet it would be difficult to bring the uncoached British ambassador into the meeting when the American ambassador would be left out. Groves' unwillingness to work through Caffery stemmed from his concern that the ambassador would not be able to keep total secrecy and might let word of the negotiations slip out; the general's intuition on such matters was strong.[19] In any case, Caffery was soon replaced by Berle, a long-time friend of Roosevelt, who was soon privy to the broad course but few of the details of the negotiations.

The rush to obtain thorium is evidence enough of the belief in the scarcity of uranium. Thorium itself appears most frequently in monazite, a phosphate containing cerium, lanthanum, and thorium oxide (thoria) in amounts ranging from 18 percent to less than 1 percent. Monazite in turn is most frequently found in sands derived from weathering of granites, pegmatites, and gniesses. Though the host rock seldom holds more than .1 percent monazite, wind and water action on sand dunes formed from this rock may provide concentrations between 3 percent and 10 percent and sometimes as high as 46 percent. The richest of the two principal known deposits in the world was in the State of Travancore at Quilon and Manavalakurichi on the Indian subcontinent. The monazite there occurred in deposits of 5 to 10 percent with the thoria content ranging between 8 and 10 percent. Thus the average thoria content of mined sands was about 1 percent. In Brazil monazite sands occurred along the coasts of the states of Espirito Santo, Bahia, and Rio de Janeiro. The thoria content of the monazite was only two to six percent. The total thoria content of Brazilian sands was therefore estimated at 5,000 tons, half that of the Indian deposits but five times the amount estimated available in the Piedmont region of the Carolinas in the United States. Some minor deposits were known to exist in Ceylon, Australia, the Netherlands East Indies, and Malaya.[20]

In May 1944 the British envisioned no difficulties in acquiring control of the Indian deposits but saw Brazil as a more difficult matter. Though thorium was of secondary importance, in that it could not be converted to uranium 233 without supplies of uranium, nevertheless the possibility of another country getting thorium to use along with a little uranium was real. Thus in May 1944 the British already had recommended that "it would not appear sound policy to exclude thorium from control if only from the point of view of preventing the cornering of supplies either from political or financial motives."[21]

Groves' reluctance to work through the existing American diplomatic staff in Rio brought about the official entrance of the department of state into the rare ores negotiations. Ambassador John Winant at the Court of St. James, who was so involved in the Belgian negotiations, had never reported on these to his department superiors; he took his orders straight from Roosevelt, whose appointee he considered himself, and from the war department. Just before Roosevelt's departure for the Big Three gathering at Yalta, Groves asked the president to let Secretary of State Edward R. Stettinius initiate the talks with Brazil. Roosevelt promptly agreed. Groves briefed the secretary and later arranged for one of his own men, Major John E. Vance, a chemist in civilian life, to accompany Stettinius to Rio. Agreement was also reached with the British that the approach made would be general, merely intended to pave the way for future talks; no definite agreement was to be reached on quantities and prices. The prime aim would be to persuade Brazil not to sell to others without the consent of the two powers and to sell to them a reasonable quantity at a reasonable price.[22] Cost was always a key British concern but not necessarily so for Groves.

It was easy enough for Stettinius to work in the matter of monazite and thorium as he talked on 17 February with President Gettulio D. Vargas of Brazil and his foreign minister, Leao Velloso. Though nominally a constitutional republic, Brazil was in fact very much under the one-man rule of Vargas. The talks primarily concerned Brazilian wartime supplies for the United States, and so one more request was not unnatural. Stettinius pointed out that the United States had bought all of Brazil's monazite production for the past five years. Though after the war the Indian product might dominate the market by virtue of its lower price, U.S. hemisphere policy implied continuing purchase by the United States of all Brazil could produce. Vargas saw no problem in providing one more strategic material. He suggested that negotiations be carried out with Valentim F. Bouças, director of the Brazilian Commission to Control the Washington Agreements on war materials. Bouças, a personal friend of Vargas, was a well-known financier, head of the International Business Machine Corporation of Brazil, and had been instrumental in negotiating a rubber agreement and the first payment on the Brazilian debt to the United States.[23]

The success of the initial approach was reported to the Combined Policy Committee on 8 March. The views of Sir John Anderson and General Groves on the raw materials situation were discussed. Both men agreed that Congo deposits should be exploited quickly, more exploration be done, and that North American reserves be conserved. Minor sources should be controlled when possible by commercial action, but no political agreements should be sought in those cases. There were some differences, as Anderson's estimate of the uranium needs for North American projects was only a third or a quarter the size of Groves'; nor did Anderson think really useful action could be taken to deny supplies of thorium to other countries. The Committee, prodded by a request from Groves as chairman of the CDT for a formal position on the acquisition of thorium, acknowledged that new experiments promised that thorium would be of great use; the cost of purchasing the prewar output of Brazil was negligible in comparison with other expenditures. Thus, despite Anderson's reservations, the group recommended that the proposed U.S.-Brazilian agreement be carried through, that the United Kingdom negotiate with the Indian state of Travancore, and that information on thorium in the Netherlands East Indies be promptly reviewed. As the United States had a network of existing trade relations with Brazil, the United Kingdom would not participate in that agreement directly, but its indirect equal participation in all rights and responsibilities obtained would be acknowledged by the United States. Travancore, as part of the Commonwealth, was not within the jurisdiction of the Combined Development Trust, but it was agreed that the United States would have a share in its supplies.[24]

Despite their auspicious beginning, the talks with the Brazilians remained on dead center. Senhor Bouças found frequent reasons to postpone any visit to Washington. William L. Clayton, an assistant secretary of state involved in intergovernmental dealings on atomic energy, and Groves finally decided that Bouças would probably continue to delay until after the Brazilian elections. As much as they preferred to negotiate on their own turf, they decided in June to send representatives to Rio. Groves' chief negotiator was a young colonel, John Lansdale, Jr., who had been a lawyer in civilian life and in whom Groves placed a good deal of trust. Though Lansdale's chief responsibility was security, he was to be used for a variety of tasks, including negotiations and the drafting of formal agreements. He was to be assisted by Joseph Volpe, another young lawyer; both men had gained experience in the dealings with Belgium the previous year. Another assistant was S. Maurice McAshan, Jr., a skilled business agent, fluent in Portuguese, who worked for the successful cotton brokerage firm of Anderson, Clayton and Company of Houston, Texas. This firm, of which William Clayton of the state department was a principal owner, had expanded its range over the years and had considerable experience in business dealings with Latin American countries. The British concurred in these arrangements, designated their own observer, and talks began in June.[25]

Agreement in principle was readily reached, but hard bargaining took place over specifics. To prevent future difficulties, Bouças thought his country should be bound not to increase taxes or official fees during the period of the agreement, but that there should be annual renegotiations of price and quantity, with quantities purchased increasing by at least 10 percent per annum. Moreover, some assurance had to be given about continuing purchases even after the Travancore market reopened.

The allied team did not want to be committed to large purchases. On the other hand, it did want Brazil to confine all exports of monazite sands to the United States or to its designated consignees, and the Americans wished the agreement to last as long as possible. In discussions among themselves, the Americans and the British representative decided that a 99-year term was too long to try for and that the initial offer to purchase should be at 1,500 tons rather than at 750. Bouças eventually agreed to adjusting prices and quantities only every three years, but soon insisted on an initial purchase of 3,000 tons.[26] The British and Americans yielded on this point as well as on a demand that they purchase sands with less than 6 percent thoria content. These would have lower prices and constitute no more than 20 percent of the total. The Americans also agreed that "due regard" would be given "the desirability of maintaining imports from Brazil" on a "fair and nondiscriminatory" basis in relation to imports from other countries.[27] This was the protection Brazil wished against future competition from Travancore. Export duties and other taxes were to be the responsibility of the exporter. In return, the allies obtained the possibility of ten renewals for a total term of thirty-three years, with no mandated increase in quantities per renewal.

As in the Belgian negotiations, the Americans—with the concurrence of the British representative—tried for a clause which would control Brazilian use of the sands by requiring prior American approval. On this Bouças "completely balked."[28] As the prime concern was use of the sands for energy production, a way around this impasse was found by wording which granted the United States right to purchase all or part of any compounds of thorium produced in Brazil "subject to the right of the Government of Brazil to reserve such reasonable quantities of these materials as may be required for normal industrial applications within Brazil."[29] Since energy production at that time was not a normal use of thorium, and as the allies were careful not to tell the Brazilians the end use intended for the material other than that it had "a potential military significance if experiments then going on were successful," such wording seemed adequate to the allies' purpose.[30] It also met the Brazilians' point of pride regarding right to their own resources. Unlike the Belgians, who knew of the energy potential of uranium, the Brazilians did not raise a similar point about thorium. The ethics of allowing the Brazilians to think their rights were protected by wording which in fact did not protect a key right was not discussed in any printed record. At the time, the future practical use of thorium for energy-related purposes was still uncertain, and the security of the allies justified secrecy.

Negotiations were completed in a remarkably brief time, if compared with the protracted Belgian dealings. The U.S.-British negotiating team made its initial contact with Bouças on 27 June and found him more than ready to proceed. Further discussions the next day led to a rough draft by late afternoon. Bouças showed it to President Vargas that evening and held a hasty meeting with the allied negotiators at the train station still later that night to inform them of Vargas' insistence on the 3,000-ton minimum purchase. While Bouças traveled, the team conferred with Berle over the non-discrimination clause and technicalities regarding completion of the agreement. Decision was made to meet Bouças' request for purchase of grades of sands under 6 percent thorium in order to improve export control.[31]

The team was reluctant to concede the point of minimum tonnage, and when meeting again with Bouças on the sixth of July used it to bargain for fifteen rather than ten renewal periods. Bouças refused to budge, for the commercial profits were of importance to himself and Vargas. The Brazilian played his ace card, stating that any concession on the points of minimum purchase or renewal options might lead Vargas to demand an entire renegotiation of the treaty by the foreign office, a matter that might take months. The allies yielded, and the memorandum of agreement was signed. On 10 July formal notes were exchanged by the two governments acknowledging the accord.

The unusual speed with which agreement was reached was linked with the unusual—or at least non-standard—channel by which it was drafted. Once again the chief allied negotiators were not members of the foreign service: McAshan and Frank Lee, the British representative, who was from the British treasury delegation in Washington and thus responsible to Anderson. They appear to have been given full authority to negotiate the best arrangement they could in the shortest amount of time, perhaps because it was thought that any prolongation of the discussion might lead to Brazilian questions, security leaks, higher prices, and interminable delays. Vargas purposely kept negotiations out of his own foreign office, and Bouças took steps to keep the agreement out of the foreign office document registry system, both to save time and to protect security. In Lansdale's view, Bouças was entirely the key to the negotiations. Even the quasi diplomatic social functions to which the negotiators were invited included few governmental officials. As McAshan and Lansdale later commented, while Bouças gave the impression of being sympathetic, he nevertheless "skillfully used the fact that the agreement was being handled 'out of channels' as a means of trading hard and driving a good bargain for his government."[32] Given the existing political situation in Brazil, the Americans thought that any detour through the foreign office might result in considerable delay, weaker export controls, and most likely a shorter life for the agreement than the thirty-three years obtained.

Brazilian Foreign Minister Velloso eventually did have to be brought into the picture. Difficulties promptly arose. The minister's pride was wounded because Vargas had not sent him direct authorization to conclude the accord but had forwarded it through Bouças, whom the foreign minister in turn refused to see. Vargas had to grant Velloso a personal interview; after that there were further delays for making translations of the document satisfactory to the ministry of foreign affairs. All this may have appeared petty to the allied negotiators at the time. Yet the attitude of the Brazilian foreign office in July 1945 foreshadowed significant problems in the long run.

Negotiations in Rio moved so quickly that Groves reported at a 4 July 1945 meeting of the CPC that the talks were "proving more successful than had been expected."[33] Lord Halifax, who later in the meeting would express concern regarding his country's future lack of raw material for experimentation, called for a subsidiary arrangement between the two great powers recording the joint interest both took in the arrangement. Though the memorandum of agreement being reached was, strictly speaking, only between the United States and Brazil so as to be in keeping with other supply arrangements, the British had been integrally involved in the negotiations. It was agreed that letters should be exchanged between Washington and London stating that the rights and responsibilities conferred by that agreement would be considered as acquired jointly and that the Brazilians were so informed. This was done by mid-September.[34]

Despite the hasty American start which had offended the British, the Brazilian negotiations had turned into a successful joint venture. Joint allied diplomacy, even if conducted by unusual channels, was working. In fact, it perhaps was working better because those who were practicing it were removed from the traditional foreign service compartmental roles. The success was all the more noteworthy because of the severe strains that were developing at the same time between the Americans and British over the issue of exchange of scientific information. Steps were soon taken for similar negotiations regarding monazite deposits in the Netherlands East Indies. Upon Groves' urging, lengthy CPC discussion of the value of other types of deposits in Sweden further resulted in a decision jointly to approach that country for a political agreement as well.[35]

Negotiations with the Netherlands

Talks with the Dutch and Swedes were conducted simultaneously and led to far different results. As early as 7 June 1945, Groves had urged that the Netherlands be approached; on the 19th Roger Makins of the British embassy proposed to Bundy that meetings with the Dutch be started in London with American participation.[36] A few days after the 4 July CPC meeting, Groves sent Major Clifford A. Taney and Major Harry S. Traynor, a veteran of the Belgian talks, to the British capital to meet with Ambassador John G. Winant. They in turn gathered with Sir John Anderson, Sir Ronald Campbell, and other Britishers to plan their approach to the Dutch. They were concerned about the confused and wary state of mind existing in the Dutch government; it might well distrust approaches made by great powers. An informal, one-person contact was decided upon; Anderson was selected because of his prior acquaintance with Dutch foreign minister Dr. E. N. Van Kleffens. The goal was a legal instrument whereby the Netherlands would control all exports of monazite and thorium compounds and not permit exports without the consent of the contracting parties. Such an arrangement of course would permit limited or no commercial gains to the Dutch; it was recognized that the United States and United Kingdom might have to agree to some minimum purchase in order to obtain a right of first refusal clause.[37]

Anderson approached Van Kleffens the same day, pointing out that there might be risks if exploitation of uranium and thorium were not controlled. The visiting foreign minister was sure his government would wish to cooperate; he of course had to consult with the prime minister and the minister for overseas territories. In a few days the Dutch put a negotiation team together which included, in addition to Van Kleffens, a scientific expert and a former minister of finance and chairman of the board of the Billiton Company, J. Van den Broeck. On 30 July Van Kleffens assured the allied negotiators of his government's agreement in principle and its concern that the raw materials not fall into the wrong hands. Secrecy might pose a problem, as international agreements normally went before parliament. He thought something could be worked out, however. Van Kleffens also reserved the right of retention of materials for experimental purposes. Since such a right had already been granted to Belgium, Anderson commented, this would pose no problem. He did not mention, however, that Article 9(b) of the Belgian agreement stated that any use of the ores for energy-related purposes required prior American and British consent.

The next day Van den Broeck, citing his country's proximity to Germany, insisted that the Netherlands be allowed to use thorium for defense purposes and be kept informed of the development of the allies' project. Here again was that sticky issue which had caused grave problems in the Belgian talks. When informed that disclosure was not possible, he suggested that his government might not enter into an agreement without provision for this information. Discussion then shifted to price and technical matters. An attempt by the allies to limit the amount of monazite reserved to the Netherlands to twenty or thirty tons was rejected by Van den Broeck, who observed that the Dutch could scarcely bind themselves to restrict use of their own resources; besides, the monazite would be kept from the hands of the enemy. Just whom he considered to be the enemy is not clear, but his other comments lead one to believe he still feared defeated Germany.

A draft accord was drawn up and reviewed the following morning by the various technical experts. The issue of disclosure was then taken up by Anderson, Winant, and Van Kleffens as they considered the proposed agreement on the afternoon of 1 August. No record of that gathering is yet publicly available. Apparently Anderson and Winant were able to dissuade Van Kleffens from pushing the disclosure issue. Van den Broeck was its chief backer; the foreign minister, on the other hand, may have been more flexible in view of the sizable borrowing needs of his country. Having withstood Belgian pressure when negotiating for the more valuable uranium, the allies were unlikely to yield on this for Dutch thorium. Anderson was willing to share some scientific information but gave way to the American demand that for security reasons none be released, and thenceforth he loyally defended that position.[38]

A memorandum of agreement was signed on 4 August; as such, rather than as a treaty, it did not require parliamentary approval. In Groves' later report of its content, there is no reference to any possible exchange of information. Unlike the Brazilians, the Dutch saw the agreement as purely political and showed relatively little concern for its commercial aspects. The allies were to purchase 200 metric tons, if available, of monazite sands per year for a period of three years. Sands with as little as 3 percent thoria would be purchased at a price of three shillings four pence per metric ton for each tenth of a percent of contained thoria.[39] Control of exports only to allied approved recipients was obtained, along with the right but not obligation to make additional purchases. The allies would also lend technical assistance if requested.[40] Only six successive options for three years each were noted. The twenty-one year duration of the maximum term of the agreement was distinctly less than that with Brazil and far less than the ninety-nine-year accords that the CDT initially favored. It was, however, better than the ten-year accord with Belgium.

The Dutch did make clear their intention of pursuing their own research. Therefore a clause was included to the effect "that if any peaceful industrial use by them [The Netherlands] is likely to involve danger to international peace and security, the matter shall be made the subject of consultation and agreement among the three governments."[41]

This was a most peculiar inclusion and represented, no doubt, compromise between the reluctance of the allies to see any nuclear research on the continent and the Dutch insistence on the right to pursue their own work if not granted access to allied research.

Negotiations with Sweden

While the talks with the Dutch were part of a carefully planned schedule of negotiations, the need to deal with Sweden came as a surprise. The survey report of October 1944 had not given prominence to Sweden as a supplier of uranium, though it had mentioned a "slight possibility" that significant sources might be found there. Within the survey group the British members considered American assessment of the value of Swedish ores poorly founded. Immediately after the survey report was made to the CDT, the British were asked to investigate the Swedish deposits of kölm, a coal-like mineral found in the lenses and seams of Swedish oil shales. They secretly obtained assistance from a Swedish geologist who, after a quiet on-site visit, compiled a report which sent Makins to Bundy's home at seven o'clock on a Sunday evening.[42] As the British diplomat and specialist on atomic energy matters put the issue in a memorandum the following morning, the new report was of "such consequence to the whole future of the project and its place in international relations" that it was necessary that the two governments at once consider taking action.[43]

Two major fields of kölm had been located. The smaller, near Narke, contained about 15,000 tons of oxide, while that at Vastergotland held as much as 70,000 tons. The significance of these figures is clear when it is recognized that the contemporary estimate of total Congolese production was about 30,000 tons of oxide. Mining the kölm would be easy but would have to be carried out on a large scale, for only one ton of oxide could be obtained under ideal processing conditions from about every 12,400 tons of shale. Costs, however, might not be any greater than at Eldorado.

The fields were so extensive they could not be destroyed in time of war. Experts had long known uranium to be associated with kölm. Once the use of an atomic weapon drew attention to uranium and other powers found the Congolese and Canadian supplies monopolized, it was certain there would be interest in Swedish possibilities. It was "therefore imperative to initiate action immediately if the two Governments are to have a reasonable chance of success."[44] If nothing were done, other parties "sooner or later" would find in Sweden ample material for atomic research and production.

Technical factors made an effort quickly to exhaust the supply by commercial contracts with Swedish firms impracticable. For that reason Makins urged a political approach to the Swedish government. He recommended that Sweden be warned that it would be dangerous to leave exploitation of the deposits to uncontrolled private or foreign groups. Rather, the government should reserve all mining rights to itself. Whether the deposits should actually be worked could be left to the Swedes to decide. To aid them in refusing supplies to another country, His Majesty's Government was prepared to make a long-term contract for purchase of a reasonable output. Makins wondered, however, if the Swedes would also demand political guarantees or participation in industrial development, as had the Belgians. Or would they trade uranium for coal, which they badly needed? Would their neutral policy prevent them from dealing only with two nations?

Word got to Groves quickly. His staff estimated that from 1,000 to 3,000 thousand tons of oxide could be extracted per year in Sweden at a cost approximately two to three times that experienced in the Congo. The state department agreed with him that a deal should be struck soon to get the fullest possible control of supplies of Swedish uranium. When Makins' proposal for negotiations was brought before the CPC in its 4 July meeting, a prompt go-ahead was obtained. For security reasons, London was considered the best site.[45] Later Groves asserted that his desire to approach Sweden had long been thwarted by British doubts of the value of kölm and by the Londoners' opinion that as Sweden was in the British sphere of influence "negotiations in which the U.S. took part would not be welcomed."[46] Yet the indications are that in fact the British encouraged American participation once the need for a deal was recognized.

Groves' emissaries, Majors Traynor and Taney, reached the British capital the second week of July to plan approaches to the Dutch and Swedes. Hershel B. Johnson, U.S. ambassador to Sweden, was summoned to London and met with Winant, Anderson, and others to plan the approach. Winant was especially concerned by the delicacy posed by Sweden's geographical location and the manner in which her form of government restricted the executive's freedom to make security-cloaked agreements. Time would be lost if Sweden were asked to send representatives to London, who would constantly have to refer home for instructions. Anderson suggested that Johnson therefore propose to the Swedish foreign minister that Sweden not permit export of uranium without the consent of the two governments and grant them the right of first refusal of all uranium supplies. In return, the two governments would agree to buy a reasonable quantity of uranium per year. Final negotiations would still be held in London. This approach seemed feasible and was put into writing. On the 20th the draft won approval from Major Vance and Colonel Lansdale, now on the scene fresh from the Brazilian negotiations, Johnson, Winant and, after a small addition, from Anderson.[47]

While stopping over in Washington in passage from Rio to London, Lansdale received a copy of a memorandum entitled "Heads for Proposed Tripartite Agreement Between the US, UK and Sweden" which was formulated in Groves' precinct, most likely by him personally. Paragraph one stated that Sweden would limit her exports of kölm or uranium to the United States and United Kingdom as long as those countries agreed to purchase all such materials produced in Sweden not needed for her own domestic use. Paragraphs two and three called for Sweden to foster the search for kölm, with the aid of American and British technicians, and to ensure execution of commercial agreements. Marginalia on a copy which remained in Bundy's files indicate these features were the prime object of the accord, "essential as to spirit" and "desirable as to details."[48]

A fourth paragraph, "not to be used unless necessary," called for the United Kingdom to supply any deficiency of coal Sweden might experience if that country's domestic supply were curtailed as a result of the effort to produce kölm. The fifth brief paragraph stated that the agreement should be effective for seventy-five years. The figure "75" was replaced by a "99," and "99 almost essential" was written in the margin.[49]

Most interesting of all was a possible additional paragraph "which Sweden might ask for." It read simply "The U.S. and the U.K. unconditionally guarantee the political and territorial integrity of Sweden." The margin comment was direct: "impossible for us I think."[50] An attached page listed principles of a commercial phase should there be a desire to include some in the agreement. All three were bland statements regarding efforts to ensure production, furnishing of capital and equipment, and pricing.

The insistence on a ninety-nine-year deal sounds much like Groves. The notion of an unconditional guarantee of Sweden at a time when the United States was hesitating over commitments even to Great Britain is amazing; it surely would have shocked the state department. If nothing else, it reflected how determined and how far some individuals were willing to go in an effort to monopolize the world uranium supply for the safety of the United States.

The author of the marginalia is not known. Perhaps they were Bundy's notations made in connection with a conference with Lansdale regarding the current Swedish negotiations. Or were they Makins'? Bundy could well have shared the memorandum with the British emissary. In any case, Lansdale and Volpe received the memorandum prior to their departure for London on the 18th. By the time the two men had arrived on the 20th the negotiators in London had already reached substantial agreement on an approach, one that matched closely the first and key paragraph of Groves' memorandum. Lansdale, however, held serious reservations as to the possible success of the negotiations. Why, after all, should the Swedes agree? They had their neutrality and were neighbors to both the Russians and the Germans, neither of whom they wished to irritate.[51] Groves himself must have realized the effort was a long shot, to have been willing to risk such extensive commitments as purchase of all Swedish uranium and an unconditional guarantee of Sweden's political and territorial integrity.

The approach to the Swedish government was initially left to Ambassador Johnson, since the British ambassador was newly arrived from a post in Argentina and had little background regarding either the Swedes or the atomic project. Johnson's task was complicated because Sweden was changing foreign ministers the first of August.

The ambassador's opening talk with the secretary general of the Swedish foreign office went well, although reservations were expressed about any commitment for "a long period of years," as the U.S. and U.K. proposal had phrased the matter of duration. Johnson took care not to mention to the Swedes the true extent or value of their deposits or how the allies had learned about them. On 3 August Osten Undén, the new Swedish foreign minister, assured Johnson there would be no difficulty over control; that could be arranged immediately. He requested that American experts visit with his own.[52] Though the British especially had wanted the actual negotiations to be held in London, they felt they should not insist under the circumstances of the Swedish request. Lansdale and Vance travelled to Stockholm, accompanied by James Sayres, a British expert who was to assist the British minister to be involved.[53]

The three worked out a draft memorandum setting forth the British and American objectives. The initial intent was to ask for a thirty-year agreement, with option for another thirty-year renewal. The Swedes were to limit exports of kölm and uranium compounds "to consignees designated or approved by the two governments."[54] The United States and the United Kingdom would purchase all uranium-bearing residues produced in Sweden, save for "such reasonable quantities of such materials as may be necessary for scientific research or other peaceful use not involving the production of energy." The Swedish government was to use "every effort to produce the greatest possible amount" of kölm. The price paid should be fair and equitable, taking into account the cost of mining. Sweden was to furnish all available information on the occurrence of kölm and uranium in its territories. Should the Swedish government determine at the time of renewal that Sweden's "interest demands the retention within Sweden for peaceful use" the materials in question, then the two governments would have the right of first refusal of any kölm or uranium left over for export. It would be

> further mutually agreed that in such event should the peaceful use of such materials result or appear likely to result in the production of materials useful for military purposes then the parties hereto will enter into consultation and agree upon the most effective means of controlling such use.[55]

The proposal obviously bore little resemblance to the 18 July suggestions prepared in Washington, and it was changed further two weeks later on the urging of London authorities who had not been party to its drafting. These—most likely Anderson—were convinced it would be a mistake to press the Swedes to develop production on a considerable scale. They held the view that "the most important objective was to secure control"; a definite obligation to purchase specific amounts should be undertaken only if Sweden insisted on it. If absolutely necessary and as a last resort, they would agree to a purchase, or preferably an option to purchase, of a maximum of 500 tons per year. Even this seemed to involve production on a dangerously large scale.[56]

A key British concern was, in fact, to keep Swedish production low. It was bound to be slow, and deposits were "so extensive that no great gain in security would be achieved, since for many years the amount left would be so great that a little more or less would not make much difference."[57] Yet abnormal activity would be a provocation to other powers.

London officials wished to reserve the right, but not the obligation, to purchase. It would be dangerous of course to allow stocks from normal production to accumulate in any appreciable manner; these should be removed. They asked that the pertinent paragraphs of the proposal be redrafted along lines similar to those of the Netherlands agreement, granting right to purchase rather than obligation and a right of first refusal of whatever stocks became available for export. Their information from Stockholm suggested such terms would be acceptable to the Swedes. Groves acquiesced.[58] As in the Belgian negotiations, it is evident that the Americans grasped for all they could get, being concerned about supplies, while the cost-conscious British desired protection from being forced to buy materials they might not need. This last was always a concern of Anderson, who seemed to have less interest in preemptive buying than did his emissary in Washington, Roger Makins, or Groves.

Meant as "a suggested form of the agreement," this first written proposal seemed to Undén to go "a long way."[59] The foreign minister again indicated willingness to control exports but hesitated over a clause, similar to one in the Brazilian agreement, that required prior consent by the two powers to any exportation of uranium or kölm. Johnson insisted, saying that the Swedish government might at a later date modify its controls in the absence of any such constraining or contrary obligation.

> We danced warily around the subject of Russia but Mr. Undén and I had a perfect understanding on this matter. He suggested I was afraid that the Swedish Government might not be able later to resist pressure for granting at some time in the future export licenses for this material.[60]

Negotiations continued between the experts and between the ambassadors and the under secretary for foreign affairs. The question of Sweden's use of her own resources was raised, but this did not take first place. Neither did the commercial issue of the amounts to be produced or purchased. The key issue was a political one, and, as Groves later ruefully commented, "apparently, their [the Swedes] actions were governed entirely by their estimate of the seriousness of Russian reaction on any agreement entered into by them."[61]

News of the Hiroshima explosion carried in the papers of 7 August made clear the importance of the discussions. That same day the Swedes raised the issue of their neutrality, suggesting they would have to refuse any purchase requests by the two governments as "they firmly intended to refuse any requests made by Russia."[62] The new allied proposal, revised according to British request, was put forward on the 22nd. On the 28th, Johnson attempted to meet the neutrality issue by suggesting that it would be to Sweden's advantage to have obligations to the United States and the United Kingdom so that any future demands by other governments could be referred to those great powers. Moreover, since the two already held a majority of the world's uranium resources, Swedish neutrality regarding the rare ore might be a dangerous temptation to others. Johnson also tried without much success to dull the issue of the legality of a secret accord entered into by the government without consultation with the Riksdag.

The ambassador recognized that the Swedish government was hesitant and not to be pushed. On 11 September 1945 Prime Minister Per Albin Hansson told Johnson that secrecy could not be maintained because of the constitutional requirement that the agreement be put before the Joint Foreign Affairs Committee of the Riksdag, which could in turn refer the matter to the entire larger body. Moreover, "it was impossible to word the agreement in such a way that it would not be regarded by Russia as a political act of unfriendly nature on the part of Sweden and would result in a further deterioration of the relations between the two countries."[63] And of course the United States and the United Kingdom could not make themselves responsible for Swedish-Russian relations. He would, however, see to it that controls were placed on the export of kölm and uranium.

Johnson discussed this rejection with Groves' representatives. They concluded there was little to be done. Lansdale thought the Swedes had stonewalled skillfully and, despite their friendly hospitality would not be budged. Vance, after numerous discussions with Swedish scientific experts and geologists, was coming to the conclusion that the kölm deposits were not as important a source of uranium as some persons in London and Washington thought. The men did decide to ask the Swedes to include in their written rejection of the overture promises to make information available to the two governments on Sweden's resources, to inform them of any other approaches, and to grant the two first refusal for the purchase of uranium should Sweden ever rescind its export restrictions.

Undén agreed only to oral promises regarding the first two points and rejected the third until more vague wording was established. This did not involve such a specific commitment but rather agreement that, should a change in Swedish policy be necessary, the two governments would be given first opportunity to discuss the impact of such a change and to reach "mutually satisfactory arrangements."[64]

The written statement of the Swedish position began by saying that the government could not, on constitutional grounds, make the secret agreement proposed. Other countries with somewhat similar constitutions did manage secret agreements, but it should be remembered that Sweden was a neutral and her executives did not have extensive wartime powers (the European war was now over anyway). If the constitutional argument seemed primarily a paper one, the next paragraph got to the heart of the matter:

> Political considerations make it equally impossible for the Swedish Government to put an option relating to uranium materials, by means of a secret agreement, in the hands exclusively of two of the great Powers of the world.[65]

Though Undén made no reference to the allies' desire for the right to agree on methods to control any future Swedish production of materials for useful military purposes, this also posed a serious political issue of national sovereignty and defense.

The defeat of the allied negotiators was far from total, however. In the Congo negotiations, their goal had been to obtain materials for their own use. This was not the case in the dealings with Brazil, the Netherlands, and Sweden, where the chief interest was preemption. As the first sentence and the marginal comment on the 18 July draft stated, and as the British also argued, the prime object of the proposed agreement was to persuade Sweden to limit her exports of kölm and uranium to the United Kingdom and United States only. The notion of limitation was at least achieved, as the Swedish government did agree to introduce legislation into the Riksdag forbidding any exportation of uranium without government consent and also any export of the materials from which it was derived. Undén believed such a law would announce to the world that Swedish uranium was for Sweden only.[66] Should a change in policy occur, the two powers would have their day in court.

Johnson considered the obligations accepted by the Swedish government to "fully achieve our essential purposes."[67] His conviction that the Swedes were sincere was bolstered by the foreign minister's promise to refuse export licenses for uranium until the planned legislation was passed—which occurred in due course. Undén had also orally agreed to inform the two powers of any serious inquiries about uranium from any other state.

Success of the Preemption Efforts

Thus by the close of 1945 substantial steps had been taken to implement the recommendations of the survey report and the Combined Development Trust. A supply accord was signed with Belgium and preemptive agreements were reached with Brazil and the Netherlands; Sweden had given written assurance she would allow no exports. All of the negotiations took place under the shadow of war and were given impetus by it. The central pact was with a state highly dependent on the allies for military and reconstruction assistance. The Belgians made the greatest contribution, and the Dutch, also in need of allied support, were cooperative on the lesser scale of their agreement. Negotiations with Brazil and Sweden, countries less affected by the war, were ticklish and in the latter case failed to produce the substantive commitment desired in Washington.

Attempts to gain access to the monazite sands of the State of Travancore on the southwest coast of the Indian peninsula had been led by the British. As the British had by treaty rights control of the state's foreign policy, it was assumed that negotiations would move quickly. Yet progress was slow for a variety of reasons, including the unsettled situation in India and probably Anderson's relative lack of interest in thorium. Export controls of monazite and thorium compounds from Travancore were, however, enacted by the fall of 1945. No thorium could be exported without a license issued after reference to the ministry of supply in London.[68] This export control was so effective that a shortage of gas lamp mantles, in whose manufacture thorium was an essential component, promptly developed; extensive correspondence finally allowed small quantities to be exported to a few mantle-producing firms.

In Portugal steps were being taken to buy out the last remaining Portuguese minority shareholders of the Urgeirica mine; the majority share acquired by the United Kingdom Commercial Corporation was bought by the Combined Development Trust in February of 1945. Further investigation of Portuguese deposits by both English and American teams resulted in the decision to buy the Transcosa concession, the Empresa Mineira, and approximately forty or so abandoned concessions. The ores were of low grade and might not be of great use to the Trust nations, yet they could be valuable to other nations attempting to put together a sufficient supply for meaningful research. No political agreements were involved. Though the commercial transactions conducted by the British were slow and often involved what they considered improperly high prices, progress was steady.[69]

It was no doubt with considerable satisfaction that Groves, now at the rank of major general, could write to Secretary of War Robert Patterson on 3 December 1945 that the Trust nations controlled 97 percent of the world's uranium output from producing countries and 65 percent of the world supply of thorium. Of the large deposits of low-grade uranium ore capable of early commercial development, 60 percent were controlled by the British Empire in South Africa and 40 percent were in Sweden. Low-grade ores requiring great recovery costs were in the United States, India, Russia, and Argentina. Uranium discoveries in China, Manchuria, and Portuguese West Africa were possible, but the industrial and technical capacities of these countries posed no challenge. Thus it appeared that "the only countries, outside of the Trust areas, having resources and industrial power which might challenge the dominant position of the Trust group of nations in the near future are Russia and possibly Sweden."[70]

The Combined Development Trust had, in little more than a calendar year, taken substantial steps in fulfilling all the actions recommended in October of 1944. Only in the matters of deposits in Czechoslovakia and Madagascar had little progress been made. The reasons were political. The occupation of Czechoslovakia by Russian troops and the manner in which President Eduard Beneš was required to negotiate with Soviet and Communist party representatives indicated the futility of any approach. As for the French-ruled island of Madagascar, the situation was touchy. On the one hand, there were Frenchmen there whose loyalty during the war had been to the Vichy regime now overthrown. On the other, British and especially American relations with the forces of Charles de Gaulle who now controlled France were scarcely smooth. The Communist political presence in France was evident, and in Madagascar many of the native population were restive and calling for independence. Then too, Joliot-Curie was trying to restart his research and could be expected to insist that whatever sources of uranium France did control should be kept for her own scientists. In short, the situation was complex, the politics unfavorable, the chance of success doubtful, and the prize of limited enough value that no serious effort to obtain control of the Madagascar deposits was launched.

Discovery of the value of Swedish resources was warning enough that survey efforts should not cease. Further updating would be called for, including another general report in 1945. On-site research was conducted in Australia, New Zealand, and elsewhere. The then known deposits in Australia were not thought worth exploiting, but investigations in South Africa were more suggestive. Earlier efforts in South-West Africa had been unproductive, and initial estimates of uranium in the gold mines of the Rand in South Africa were low. But George Bain was convinced that these had to be wrong, and in time he would be proved correct.[71]

The CDT was aware of the uranium at Joachimstal and apparently did not think the Russians would be able to mine there extensively enough to gain the needed amounts of oxide. Despite the survey efforts of the UMDC and the Murray Hill area, however, the CDT did not know of the valuable deposits in Saxony, just north of the East German border with Czechoslovakia and the Joachimstal mines. Discovered by the Germans in 1943, the deposits were explored by the Soviets in the months after June 1945; mining operations began a year later and were feverishly expanded after April 1948.[72] The existence of these Saxony deposits were to vitiate, though not render completely ineffective, Groves' massive preemption effort.

The preemptive bargaining differed markedly from the approach used in the Belgian negotiations. Speed was a priority, and in order to achieve a better pace the U.S. and U.K. negotiators were granted increased autonomy and exercised greater flexibility. Groves' forceful plunging ahead did cause some difficulties with the British. Yet the good working relations his representatives established with their British counterparts helped smooth this problem, and the speed was of value. The more delays there were, the more likely negotiations would not be complete prior to the first utilization of the bomb. Once the bomb was detonated, then the willingness of other countries to control exports or sell uranium at modest prices might be less forthright. The extent to which the explosions at Hiroshima and Nagasaki influenced Swedish thinking is not clear, but the posture the British estimated in early August to be favorable to a written agreement had moved by September in a negative direction.

Representatives of the two powers skirted the issue of energy research whenever possible; the debate with the Belgians had been too protracted and uncomfortable. Moreover, if all that was being asked was export control, then there was less reason for the other countries to be taken into the confidence of the great powers. When the energy matter was raised, the allies' concern for security was such that their insistence on overseeing and controlling the energy research of nations such as Sweden must have given pause to small-state diplomats naturally alert to protect their country's sovereignty from the pressures of great powers. The British and especially American insensitivity in this regard did not pose too great a problem in the preemption agreement negotiations, primarily because the issue came up in its most pointed form only with Belgium, the one power the allies thought might have the research ability, industrial potential, and interest to make progress toward military applications.

Though Groves and the scientists never ceased to warn of the possibility, indeed surety, of Soviet experimental progress, his efforts to gain preemptive control of uranium and thorium supplies demonstrated the conviction that foreign research could be substantially slowed. The extensive debate in 1945 (which has since continued) over whether President Harry S Truman should inform Premier Joseph Stalin of the bomb during the Potsdam Conference in July 1945 suggests that many high-level Americans were not sure the Soviets knew how far along the Americans were and, most importantly, whether an atomic bomb could be made. The preemption agreements thus most served the Western great powers by providing a false sense of security which would make the shock of the Russian early detonation all the greater, provoking a reaction that would only lead to greater Cold War tensions.

Beyond this, the three sets of negotiations are interesting measures of the extent of British-American influence. Brazil complied quickly through unofficial channels with little concern about parliamentary regulations, the negotiations greased by the commercial gains involved. The patterns of dollar diplomacy were hard to break. The Dutch viewed the matter as strictly political and cooperated readily but with careful questioning. Sweden offered the gesture of voluntary export control, but refused any action which might endanger her carefully balanced position between the West and the U.S.S.R. Above all, the negotiations underlined the significance of the Belgian accord and of the confidence and commitment of men like Sengier and Spaak to the Western cause.

4 · Price, Politics, and Pride: Further Negotiations with Belgium

Several of the ore procurement accords reached during the war and immediately following its close in Europe contained promises which lacked clear definition. In the case of Article 9 of the agreement with Belgium, specifically designed wording papered over differences of perception or desire which, had they been pursued to their full development, might have forestalled any signature of an agreement at all. The will to agree had been present, however, and the working out of the arrangements was left to the future.

Yet it could not be postponed indefinitely. The Congo was by far the allies' best current supply of ore. Moreover, the issues carefully cloaked in Article 9 were relevant not only to relations with other ore-producing countries but also of signal importance for continuing U.S.-U.K. scientific relations and to American domestic concern regarding proliferation of nuclear knowledge and weapons. Then, too, any agreement with Belgium regarding price would, of course, have an effect upon all other contracts.

The Belgian diplomats, perhaps as much or more than the British and Americans, wanted to keep their uranium out of "the wrong hands." They also wanted more specific things: a good price for their ore, removal of tariff barriers against their radium products, some uranium for their own ceramics industry and scientific research, and participation in the commercial and industrial applications of the military research undertaken by the two great powers. From the tenor of certain discussions, although explicit evidence is not available, it seems possible that the Belgians also hoped that agreement regarding Congo uranium would bring with it Anglo-American support for Belgium in her postwar rule of the Congo. The British had, after all, been critical of Belgium's colonial enterprise since the time of Leopold II's excesses; American and especially President Franklin D. Roosevelt's negative attitude toward colonialism was well known. In any case, the British and Americans were not reluctant to play upon the issue, as they pointed out how entry into agreement with the United States and the United Kingdom would help protect the Belgian national patrimony.

Not entirely resolved to the satisfaction of all concerned at the time the Tripartite Agreement was initialed, the issues of price and scientific collaboration proved difficult of solution over the ensuing years. Another issue which provoked extended debate was the secrecy of the accord. Article 10 stated that the memorandum of agreement "should be treated as a military secret in keeping with its purpose."[1] But there would be many Belgians, aware to lesser or greater degree of the Congo uranium deposits, who would want to know what was being done with the ores. Some of the curious would sit in the national chambers and ask questions with significant political overtones.

Monetary Profits and a New Tax

The contract between the Combined Development Trust and the African Metals Corporation signed in September 1944 dealt only with delivery of an initial 3,440,000 pounds of uranium oxide. Groves knew that further research and eventual production of weapons, even in the relatively near future, would be hampered unless more shipments could be purchased. The Combined Policy Committee agreed that there should be no delays. In endorsing the recommendations brought before it by Sir John Anderson at its 8 March 1945 meeting, the Committee accepted his opinion, given the Trust's view that North American uranium supplies should be conserved, that efforts be made to determine the ultimate resources in the Belgian Congo, and

> that the deposits in the Belgian Congo should be exploited as rapidly as possible and the material, both of high grade and low grade, [be] removed to safe territory. . . .[2]

Groves, Sengier, and their representatives then settled down to six months of bargaining. Complications arose from uncertainties regarding the depth to which high-grade ores ran, the variability of extraction costs depending upon the depth from which the ore was mined, and the fluctuations in currency value that might be anticipated over the period of the contract. On 27 October 1945, two separate contracts were signed by the Trust and the African Metals Corporation. One of these arranged for the purchase of all the uranium oxide contained in high-grade ore mined down to the 150-meter level to a maximum of twenty million pounds and for an option on all oxide in lower grade ores mined to that point. The other contract covered the purchase of all remaining oxides which could be economically mined at Shinkolobwe after delivery of the twenty million pounds, yet still within the ten-year option period of the Tripartite Agreement.

According to contemporary estimates, the 3,440,000 pounds of uranium oxide originally contracted would be delivered by March 1946, at which point the ten-year period would go into effect. The next twenty million pounds of unprocessed ore were expected by 1949, and a final amount of approximately forty million pounds of ore was anticipated prior to March 1956. Radium and other precious metals contained in the ore belonged to African Metals and would be returned to the corporation in the sludge resulting from the processing of the ore for uranium oxide.[3]

Price remained a difficult issue, complicated by questions regarding the actual costs of production and whether different prices should prevail for different-quality ores. It was finally agreed that high-grade ores should bring a price of $1.90 per pound. For ores having an oxide content between 25 percent and 5 percent, the price would be $1.35 per pound, while ores with a content of less than 1.5 percent would bring only fifty cents per pound. The Belgians, aware of what inflation could do over a ten-year period, asked that the price be guaranteed in gold. This the British could easily do, as their currency had gold backing. Not so for the Americans, who therefore had to agree to a premium of 15 percent on their share of the purchase. Letters were exchanged between the two English-speaking powers, assuring that while the actual price paid by the Americans might be different from that paid by the British the two countries were assuming equal obligations as envisioned by the terms of the Combined Development Trust.[4]

Profits for Union Minière under these contracts were substantial. Whatever the costs of production, they were low relative to mining elsewhere. The ore was accessible by stripping, much needed equipment was provided by the Americans under the 1944 agreement, and wages paid to native laborers were low. In the eyes of the American ambassador to Brussels in 1946, the Belgians were "getting handsome recompense."[5] A later ambassador would argue, however, that "Sir Edgar supplied his uranium at very reasonable cost. He was as eager as anyone to defeat the Nazis."[6]

The amount of profits accruing to Union Minière concerned Sengier. Questions were bound to be asked. Disclosure might bring assertions that such profits belonged to the nation as a whole or that they be reinvested in the Congo colony rather than distributed to Union Minière stockholders around the world. Already in December 1945, Paul Libois, a Communist deputy in the Belgian chambers, proposed a bill nationalizing uranium deposits in the Congo and Ruanda-Urundi. Nor was Communist pressure the only potential source of danger: Belgian officials responsible for the Congo might also prove difficult.

Sengier and one of his lieutenants, Van Bree, considered making a gift to the Congo colony from the uranium profits. They saw such a gift as a social and political rather than a business matter, designed to stifle Communist or Russian inspired criticism "and to insure the complete cooperation of the Minister of Colonies and the Governor General of the Belgian Congo in supplying uranium to the United States."[7] In the fall of 1946 during a luncheon meeting of Union Minière directors, Van Bree therefore offered Governor General of the Congo Maurice Godding a check for 100 million Belgian francs for social welfare work in the colony. Godding declined the offer because he did not wish to disclose the exorbitant scale of profits derived by the company from its uranium contracts. On 29 September, however, Godding discussed with Spaak and Minister of Colonies Camille Huysmans the Union Minière monopoly and large profits. The three felt that such huge profits accruing to a private corporation might well be criticized and therefore should be directed to welfare work in the colony. But it was important to have a plan for this.[8]

As revenues poured in, Spaak and Godding became increasingly concerned and soon contemplated the institution of a license for the exportation of uranium, the revenue from which would be used for welfare in the Congo. Sengier assured U.S. Ambassador Alan Kirk that the company, not the United States, would have to pay the fee and that he would be sure not to let it jeopardize the contract. Godding did not object to the Union Minière contract but needed window dressing to satisfy chiefly Communist criticism. In any case, Kirk thought Spaak to be the controlling personality.[9]

The United States ambassador answered Sengier that his superiors saw export licensing as purely a Belgian internal revenue matter and had no objection. Yet when Godding called at the end of October 1946 to indicate that another interpellation was anticipated from a Communist deputy who alleged that the price being paid was too low, Kirk asked him not to divulge the figure. In the absence of an established world price, any figure could be criticized. Godding obligingly stated in the chamber that uranium was being sold at a fair value, with the price varying according to the richness of the oxide content—a correct reply which dodged the issue of an exact price.[10]

The possibility remained that Congo ores might be nationalized. In May 1947 Sengier reminded Spaak that since the Union Minière leased its extraction rights from the Congo government, nationalization already existed in fact. Governmental recapture of some of the profits was acceptable to Sengier, but only if the proceeds were used to benefit the natives rather than to balance the Belgian budget.[11]

If the executive's attitude toward a tax increase was relaxed in May, it changed abruptly the following month. Pierre Wigny, the minister of colonies in a new cabinet formed by Spaak in March, finally came forth with a much-publicized program for the development of the Congo. Sengier questioned the positive effect of Wigny's program for the colony and especially the size of the proposed new tax on uranium exports—60 francs per kilogram on top of the existing 6 percent advalorem tax. Sengier considered the amount excessive; it took most of the Union Minière profits. Were it not for the special importance of uranium he would be tempted to stop extracting it, he commented. Still, the tax was better than nationalization. Sengier thought he had won a significant victory in insisting that all the tax monies be kept in the Congo, though he suspected it probably would not all be spent on the natives. While Wigny was a member of the moderate Christian Social party, he held leftist tax ideas, according to Sengier, who attributed the size of the tax to Wigny rather than to Spaak.[12]

The new royal tax decree was promulgated on 9 June 1947. It stated that because of the war no decision had been reached until recently regarding fiscal regulations on uranium. The 60 franc/kilogram tax would be levied retroactively on all oxide exported since 1 January 1944. Ceramics industries were, however, protected from possible price increases or retroactive payments by an exemption for up to 200 tons of ore per year shipped for industrial rather than energy-related purposes.

Sengier did not protest too loudly, for fear of stirring more trouble. He even arranged for Union Minière to donate ten million francs each to the Universities of Brussels and Louvain for physics research in order "to keep Libois and his friends quiet."[13] When the American chargé in Brussels asked him how long Sengier thought twenty million francs would keep Communists quiet, the executive retorted that "it would probably not keep Communists quiet at all but should have healthy effect in Belgian scientific circles and tend to lessen their susceptibility to Communist arguments."[14]

Wigny's success in pushing through his tax proposal was related to complex currents in Belgian domestic politics at the time. The move also reflected the attitude of a succession of Belgian cabinets that something had to be done in regard to the uranium deal to assure the Belgian populace that the nation's patrimony was not being squandered. This inclination was further strengthened by pressing economic needs and by the frustration the cabinets had experienced with certain terms of the 1944 agreement, namely its secrecy and the matter of industrial uses of uranium. Any future actions regarding taxation were to be greatly influenced by these issues. Moreover, as the Union Minière profit margin had been so sharply cut by the size of Wigny's tax, any further tax increase would not be solely an internal Belgian matter but would necessarily spill over to affect the African Metals Corporation's agreement with the Combined Development Trust and thus Belgium—United States—United Kingdom relations.

Maintaining Secrecy of the Terms of Agreement

Given the Combined Policy Committee's hopes to control as much high grade ore as possible, to extract the Congo ores "at the earliest possible moment" so that "the Belgian Congo deposits would be worked out at the end of ten years," it is not surprising that American officials were disturbed by the text of a bill proposed in the Belgian chamber in December 1945.[15] It stated that all uranium deposits in the Congo and Ruanda-Urundi were reserved for the state, required that deposits currently controlled as private concessions be taken over by the state, and prohibited the holding, transportation, or transaction of ownership of uranium-bearing ores in Belgium, the Congo, or Ruanda-Urundi without the authorization of the Belgian minister of finance or the governor general of the Congo.[16]

The Americans were therefore relieved when Foreign Minister Spaak assured them that the bill would not even reach committee prior to dissolution of the chambers and a new election. But the strength of the left in the ensuing campaign concerned Union Minière officials. The Communist party was making nationalization of uranium a plank in its platform, and it was possible the bill might be reintroduced with Socialist and Communist support. The company leaders thought it would be well to enlist the support of Spaak and Achille van Acker, the previous prime minister. These men knew the full story and, although Socialists, they were not doctrinaire; they might be sympathetic to the Union Minière and American viewpoint.[17]

A Socialist government led by Spaak lasted only a week, and eventually a Socialist-Liberal-Communist coalition cabinet led by Van Acker with Spaak again as foreign minister reappeared. The feared bill was promptly reinstated by leftist deputies, but Van Acker assured Ambassador Kirk that "such bills were designed for home consumption only and would not be brought to a vote."[18] Shortly thereafter, in a meeting with one of Kirk's staff and in the presence of Van Acker, Spaak indicated that Minister of Colonies Godding was formulating his own bill to supersede that proposed by the Socialist deputy Anseele. It would be less comprehensive and could by its nature be enacted by royal decree, thus avoiding the issue of parliamentary approval. Yet Spaak remained concerned that even if a decree sidetracked the nationalization bill, it nevertheless might provoke an interpellation which could be embarrassing if pointed questions were asked about the past disposition of uranium ores.[19]

Thus even the decree route was not entirely safe. Spaak and Godding brought pressure on Anseele to allow his bill to be stifled in committee.[20] Yet questions were still being asked, for example by E. Lalmand, the Communist minister of food. Though Spaak successfully evaded the issue in cabinet meeting, he was anxious about further incidents and felt "temporizing could not be prolonged indefinitely."[21]

Kirk was sympathetic to the difficulty in which the Belgian government was placed. The 1944 agreement was supposed to be kept secret, but there was much popular interest in uranium which might lead the public to put fictitious or enhanced value on the resource. The Belgians had promised not to say anything without consulting the British and American embassies. Kirk wanted to be prepared for the possibility of the disclosure of the wartime agreement and reported to the state department:

> I feel that at some point it will be best policy to be honest and with concurrence of both Belgians and British to announce frankly what has been agreed to and why. . . . The disclosure of our agreement may be by events and I suggest we should have a definite plan to forestall criticism. . . . (I)n any case I recommend the prob-

lem be studied to determine (a) what degree of secrecy is still required and (b) the question of timing if any announcement is made and by whom.[22]

Groves discussed the recommendation at some length with Sengier, whose cooperation and attitude he considered of great value. Groves' commitment to Sengier was nearly total, and he strongly believed that Sengier deserved recognition for his assistance to the United States. On 1 April 1946, Groves nominated Sengier for the Presidential Medal of Merit, mentioning his "untiring efforts" which increased production rates of uranium: "Mr. Sengier's sound judgment, initiative, resourcefulness and unfailing cooperation contributed greatly to the success of the atomic bomb project."[23] The medal was approved and presented in a private ceremony because publicity still had to be avoided. Recognition was also given Sengier by the British. In addition to being a friend and a reliable professional colleague to Groves, Sengier was also the American's single best supply contact. Even with the estimated 7,700 tons of high-grade reserves still available in the Congo, Groves feared there would not be enough uranium to supply American requirements over the next three or four years.[24] The general's viewpoint was that "great weight" should be placed on Sengier's recommendations; he saw the matter more from the position of the businessman than did the diplomats in Brussels, who were more directly concerned with the difficulties of the Belgian ministers.

Sengier believed that disclosure would be a mistake, for it would provoke "a movement to abrogate existing agreements both political and commercial."[25] He also wondered if any announcement might jeopardize current U.S. efforts to place atomic energy control under the United Nations. He did not think Spaak would want to say anything that would "muddy the waters" during this critical period.

Groves supported this view and sent a draft statement for Kirk to discuss with Spaak. Its main thrust was that the proposals which the U.S., British, and Canadian governments would soon be taking to the United Nations depended greatly on existing circumstances. Belgium should not do anything to upset the proposed plan for an Atomic Development Authority under the United Nations.[26] The British agreed with the state department that, pending clarification regarding UN action on atomic energy, Spaak should not make any disclosure. When this was explained, Spaak attempted to delay the anticipated interpellation.[27] Only partially successful, he was soon asking Kirk for guidance on what could be said.

If there was some validity to the argument regarding the United Nations, it was also something of a smoke screen. Kirk and his British counterpart in Brussels argued that the various proposals emanating from London and Washington did not meet the Belgian foreign minister's need. A more direct response was necessary, one that acknowledged that an agreement had been reached during the war, that the governments were currently content with it, and stating that when a suitable arrangement on atomic energy control under the United Nations was reached, the agreements could be reviewed.[28]

This was the course taken. Spaak, Van Acker, and Godding met with Kirk and British Ambassador Hughe Knatchbull-Hugessen on 18 May 1946. They formulated a reply to a Libois interpellation that emphasized that the regime applied to uranium during the war was similar to that for other strategic materials, that the Union Minière contract had government approval, that material for Belgian research could be made available, and that the Belgian government was considering legislation on uranium but was awaiting the decision of the atomic commission of the UN.[29]

The response to the interpellation went smoothly and relieved the building pressure. Yet Libois did not retreat, and he later proposed the creation of a committee on uranium made up of deputies and professors. He had a limited following, so the proposal was easily turned aside on the grounds that there was a subcommittee of the cabinet dealing with the matter. Nevertheless, Godding soon suggested that a Belgian Commission on Atomic Energy, made up of noted scientists and headed by a prominent individual, might be useful. U.S. Under Secretary of State Dean Acheson did not object.[30]

Soon questions arose about price, followed by pointed inquiries about Belgian participation in the benefits of atomic research. The latter were stimulated in part by Joliot-Curie. He was currently director of the French atomic energy program and a member of the French delegation to the United Nations Atomic Energy Commission. In these positions he could promote his view that pooled European talent and resources could certainly match those of the Americans and that Belgium's contribution could be uranium.[31]

Though Belgian scientists were not necessarily interested in working with Joliot-Curie, they did begin to ask cabinet members why Belgium should not be undertaking research in atomic energy with her uranium. By January 1947 pressure had mounted to the extent that Kirk saw Spaak as "really worried." Some enlargement on the statements of the previous May were needed which might "restrain debate and keep discussions from becoming embarrassingly frank."[32]

The American and British ambassadors therefore drafted another statement which covered the same points as the previous one and concluded that the arrangements "fully protected [the] natural and legitimate interest of Belgium as regards supplies which she might require for her own purposes."[33] No mention was made of the huge size of the contracts signed, the long duration of the accord, the exclusivity granted to the two great powers, or the limitation put on Belgian use of uranium for energy research purposes. Each of these matters and especially the last would have produced negative reactions in the chambers and in the Belgian scientific community.

Nor was mention made of the provision for the two governments to allow Belgium to participate in utilization of uranium as a source for commercial energy. Yet this was the area in which Spaak was now making sharp inquiries. Belgian scientific circles wanted to know if Belgium should be doing energy related research with Congo uranium. Could Belgian scientists come to the United States to participate in developments there? The Belgian war department was also asking about sending scientific attachés and exchange professors.[34]

These requests were all the more embarrassing to the American diplomats because of the passage by congress in the summer of 1946 of Public Law 585, the Atomic Energy Act, or the McMahon Act as it was frequently called for the Connecticut senator who was its chief proponent and chairman of the special senate committee on atomic energy. That act specifically proscribed exchange of information with other nations on the use of atomic energy for industrial purposes until congress declared by joint resolution that effective and enforceable restraints had been established against its use for destructive purposes. The stormy history of this legislation is complex and reflects American fears of the Soviet Union and of insufficient security efforts by other powers. The law itself clearly posed a problem for the information exchange established between the United States, the United Kingdom, and Canada during the war—more about this later—and eviscerated Article 9(a) of the Belgian agreement.

At the beginning of February 1947, Spaak pressed anew for an agreed-upon statement and for association of Belgian scientists with American research on utilization of atomic energy for industrial purposes. The government, he thought,

> will certainly be confronted shortly either by Parliamentary interpellation or draft laws. Certain members of the Chamber will more than probably demand that uranium deposits be nationalized or placed under strict control.[35]

Secretary of State George C. Marshall would not give way. He urged that the very limited draft statement proposed by the ambassadors be adhered to closely. He also instructed Kirk, if Spaak pressed his request regarding the scientists, forcefully to inform him that the Atomic Energy Act made compliance impossible.[36]

On this, the Americans differed from the British. The British also wished to reject the Belgian request, but took as their pretext a careful reading of Article 9(a). The wording there referred to Belgian participation in the *utilization* of uranium for commercial purpose, not to participation in the research necessary for such utilization. Thus Belgium could be denied involvement. Marshall concurred in the interpretation but did not wish to use it, primarily because the Atomic Energy Act was also preventing the United States from providing the English with information. In the view of Edmund Gullion, one of Acheson's assistants in the office of the under secretary of state, this was the reason why the British were unhappy with the U.S. reply to the Belgians. In any case, the British did not force a change in the U.S. posture, though they thought it "might cause the Belgians to question our [the Americans'] good faith in making the Belgian accord."[37]

Concessions Grow Necessary

Tensions over the uranium question and the participation of Belgian scientists grew sharper as a result of a March 1947 ministerial crisis. In that month the Belgian Communists pulled out of the coalition government. Spaak replaced Camille Huysmans as prime minister and put together a new coalition with the Christian Social party. It was the first lasting ministry since the war in which the Communists had not participated.

Though the ostensible reason for the collapse of Huysmans' cabinet was Socialist-Communist differences over coal prices, the U.S. embassy believed that the Communists had departed upon instructions from Moscow. This appears to have been the case, as Communists quit cabinets across Europe as a part of a shift of Soviet strategy. But Kirk and his staff were convinced that the Communists, now out of the Belgian government, would be "free to attack Government handling of uranium which is what we have feared for a long time."[38]

To the Americans, the developments were all the more disconcerting because Spaak's Socialists and the Christian Social party, in order to prevent the left wing of the Socialist party from following the Communists out of the government, had both acquiesced to a political platform plank calling for nationalization of uranium deposits. Nationalization was seen by the Americans as "contrary to our interests inasmuch as it would presumably be more difficult for a national company to refuse Russian requests for uranium than for a private company."[39] The African Metals contract might be made public, thus revealing the extent of the allies' supplies and needs, and nationalization would, of course, circumscribe Union Minière profits and attenuate Sengier's welcome spirit of cooperation.

The British and Americans were also disturbed because the new government had decided to allocate ten million francs for uranium research, which would divert some of the raw material away from the United States and Britain. The amount diverted might not be significant, said Gullion,

> nevertheless, from our point of view it is dangerous that the principle of diversion should be established and that research should be brought into Europe at a point where its materials and results might become easily available to the Russians.[40]

The dispersion would be in addition to amounts ranging between 50 and 70 kilos of ore or uranium salts given by Sengier to five Belgian universities the previous October. He had repulsed a request from Joliot-Curie for 100 tons of ore on the grounds that it was not available.[41]

The Union Minière executive had supported these universities in part to ease pressure on his firm but also because it was a natural consequence of earlier actions. At the war's end a National Scientific Research Fund had been established by the directors of the Société Générale, among them Sengier. The fund in turn had established a scientific commission headed by Professor Van den Dungen of the Free University of Brussels and including representatives from five universities. Only one of those men, a Professor Geheniau, was a known Communist.

Any drain of materials, no matter how small or politic, was anathema to Kirk. On 10 January 1947, he had written to his long-time friend Acheson that

one of my principal tasks here is to prevent any disruption of existing supply arrangements and in this I have felt obliged to discourage agitation for diversion of raw material from the existing channel, except as provided for in the contract.[42]

The ambassador, a former navy admiral, was strongly concerned about the Russians. A few weeks after writing this letter, he told Acheson that he thought the Soviets' objectives would be twofold.

One would be termination of the present contract. This would require that its terms be made public and made to appear objectionable to Belgian public opinion. The other would be to have uranium made available to socialists whose work would be made known to the Russians.[43]

He surmised that the Russians were getting their own ores from internal deposits or Czechoslovakia but that Joliot-Curie's work could still be of importance to them and that therefore pressure would intensify on the second issue. Kirk remained puzzled that Moscow had not pressured the Belgian Communists to raise a controversy over it. Spaak himself was amazed that the Russians had not tried to buy stock in the Union Minière.

Perhaps the Russians were too occupied with developments in Berlin, Czechoslovakia, Turkey, and Greece. Or maybe they did not wish to embarrass the Belgian Socialists too greatly, for as long as they remained in office the royal question was sure to bubble: Spaak was a leading opponent of any return to the throne by Leopold III. It is doubtful the Soviets were uninformed regarding the Tripartite Agreement of 1944 and current American and British concerns, given the activities of Donald Maclean, their agent in the British embassy.

Perplexed as the British and Americans were by the Russians' lack of aggressiveness in this instance, their nerves were thoroughly jangled by the Belgian cabinet change of March 1947. Reassurances were forthcoming from both Spaak and the office of Charles, the brother of Leopold III, who was serving as regent until some decision was reached regarding Leopold's return and possible continued reign. They argued that the nationalization plank had been placed in the political platform to satisfy those who knew little of the subject, but the matter would be administered by those who knew a great deal. Some formula of government control would be worked out to satisfy popular demand without altering the supply arrangement. As for embarrassment if Russia asked for uranium, it would not make much difference whether the company were nationalized or not.[44]

Even as he eased the Americans' minds on this score, the new prime minister asked them again for involvement of Belgian scientists in American research. Spaak tactfully pointed out that the United States could hardly object to Belgium starting her own research since the McMahon Act prevented her scientists from getting the information they needed.

Spaak maneuvered well, and Gullion and Kirk saw the need for a new response by the British and American governments. They were further concerned by the replacement of Colonial Minister Godding, a Liberal party member with fairly conservative views, by Pierre Wigny, a Catholic Social Christian. Four months earlier Kirk had fretted to Godding about possible changes in the position of governor general of the Congo. Godding had warned that a leftist governor general might lead to a leftist minister of colonies. This Kirk scarcely desired; he wrote Acheson that "our long term interests [are] best served when personalities concerned in Congo matters are kept to the minimum, certainly for [the] next few years."[45] Yet now there appeared a new figure who, even if Spaak assured he would "not be particularly 'in the picture,' " was a kind of Socialist.[46] A few months later Wigny's tax law which so annoyed Sengier would give concrete object to Kirk's worry.

Belgian political developments reinforced the concern of Kirk and Gullion for the procurement program. Some solution to the restrictions posed by the McMahon Act was urgent. If possible, Kirk's suggestion of bringing Belgian scientists to America should be adopted, despite the previous reluctance of the Atomic Energy Commission, the group which had replaced the Combined Development Trust in January 1947 as a result of the McMahon Act.

It is true that we are to some degree closing the barn door after the horse is stolen, but we might be able to keep some control of the Belgian research program by bringing scientists here.[47]

On April 15 Kirk further cautioned that any opposition to growing Belgian interest in nuclear research might bring the 1944 agreement into question and "jeopardize renewal of at least its monopolistic features." Dependence on "warm Belgian goodwill for our raw material" meant the question could not be ignored; the "best course is to try to channel Belgian interest to our advantage." Could an American scientist be sent to lecture in Belgium?[48]

A month later Spaak formally addressed Kirk and the state department in a note asking if full revelation of the 1944 agreement was possible. Most of its terms were known anyway except its duration. Didn't the governments exaggerate the secret aspects? The Belgian public

> probably believes that [the] secrets are much more complicated than they are and in my opinion present mystery is likely in long run to do much more harm than good.[49]

Spaak then pressed for a specific interpretation of Article 9(a). He had been informed that the United States and the United Kingdom were working on industrial utilization; how did they plan to fulfill their promise?

The American Atomic Energy Commission and the state department were indeed embarrassed by the McMahon Act. The prime purpose of the act was to "assure the common defense and security" of the United States. Under Secretary Acheson cabled Kirk that the state department was attempting to determine if that purpose gave it an excuse "to use information as a counter in the dealings necessary to assuring continuing supplies of raw materials in the required amounts."[50] It was a clever gambit, arguing that the restrictive law gave justification to release of information so as to assure national security. No decision had been reached, and some delay could be expected.

Acheson recognized "the dangers to our procurement program and to security presented by recent political changes" in Belgium.[51] He instructed his representative to assure Spaak that Washington was aware of his difficult position and was searching for a way out. Spaak should be told that use of atomic energy for industrial purposes was not nearly as imminent as Spaak believed and that the American and British governments were studying how to implement Article 9(a). The under secretary also informed Kirk that an exchange of scientists was being explored. But, he wondered, were the Belgian people aware of the danger of Communist infiltration and of the activities of the recently exposed spy ring in Canada?

Marshall and Acheson both understood, as Marshall put it, "the necessity of not embarrassing Belgium vis-à-vis the Soviet Union and the Communist opposition."[52] This was one reason they held to secrecy. Spaak preferred openness, and, while he acquiesced to continued silence, he made clear that he wanted to reveal the substance of Article 9(a) in order to assure the Belgian public that its interests were protected. He also seemed sensitive to U.S. concerns regarding security, offered to create an entirely new scientific commission, and promised that scientists sent to America would be carefully chosen. What he wanted was "not secrets but results."[53]

The American chargé in Brussels, Theodore C. Achilles, soon reiterated the need to inform the Belgian populace about Article 9(a). Publicity on this would persuade the Belgians that, due to the great obstacles involved in commercial utilization of atomic energy, they were better off to let the United States do the research. It would also "spike Communist efforts to encourage both Belgian research and opposition to giving us the whole supply"and would strengthen Belgian opinion toward America.[54] He also reminded his Washington superiors that, while Spaak could be considered sound, half his party had been against joining the new government without the participation of the Communists.[55]

The chargé had taken responsibility for communication with Spaak because Kirk was in Washington. He met there on 4 June 1947 with the members of the Atomic Energy Commission (AEC) created by the new law and with Edmund Gullion, who had become the state department's sole specialist on atomic energy matters. Kirk assured the gathering it could count on Union Minière and the Belgian government to stand by the contract, but that the Belgians thought of the agreement more as a treaty than merely an accord, a treaty that involved the good faith of the United States. What was needed was a clear statement assuring that Belgium would participate in atomic industrial energy and that progress in that field was much slower than originally anticipated.

The members of the commission for their part pointed out to the ambassador that

the Belgian government does not appear to be aware that our reluctance to reveal anything about contracts arises from the fact that we do not want the Russians to make use of the information in a propaganda drive in the UN. . . . [For the time being] we should refrain from making any statements which might give ammunition to the Russians for a charge that we were acting unilaterally, or in bad faith.[56]

The discussion was wide-ranging. Members agreed on the value of sending a scientist to Belgium to explain that industrial uses of atomic energy were still far off, despite what articles in the Sunday Supplements seemed to suggest. Ways to interpret the McMahon Act so as to allow exchange of information were being explored. Unfortunately, processes used to develop industrial energy were thus far in many ways identical to those used to develop a bomb. If some way could be found for keeping the uses distinct, it lay in the future. Meanwhile, radioactive isotopes useful for medical research might be sent as a public relations move.

That the state department took all these discussions seriously is reflected in the meetings held by Kirk with the Atomic Energy Commission, Marshall, and President Harry S Truman. Acheson was perturbed by the Soviets' filibustering and failure to come to grips with the American atomic energy proposals in the UN. The Americans concluded that the Russians wished to strip the United States of atomic weapons while avoiding any international agreement on inspection. The United States wanted the world to see clearly where the responsibility for failure of the UN negotiations lay. Therefore the Belgian agreements should not be published; otherwise, the Soviets might accuse the United States of secret unilateral self-serving while publicly trying to promote multi-lateral control of atomic energy. It would thus be a "grave error" for Spaak to reveal anything about 9(a) or the rest of the agreement.[57]

Kirk explained this to the Belgian prime minister when he returned to Brussels. Spaak was responsive and willing to "play ball."[58] That he did, for, when baited by Communist senators in July, Spaak produced a statement similar to that suggested by Kirk some weeks earlier. To shouts that Belgian uranium was being hoarded and the United States was accumulating reserves, he cooly replied, "I have made my statement and you can draw conclusions you wish."[59]

Though Spaak weathered the interpellation, it was clear that his situation was becoming difficult. American and Belgian officials alike saw Communist pressure within Belgium as growing, and more than a few feared the possibility of war. Acheson and Kirk were quietly disseminating information to the Belgian public about Communist espionage in Canada and elsewhere. The under secretary saw political developments in the small country as adverse to American interests, and Sengier was deeply annoyed by Wigny's new tax.[60]

Increasing monetary gifts and small amounts of uranium could not be expected long to quiet the Belgian scientific community. Nor was the argument that UN debates required secrecy on the agreement going to have lengthy validity. The UN Atomic Energy Commission was scheduled to report in September, and the organization itself expected to reach a decision on the American proposals soon afterward. If United States insistence on secrecy were keyed solely to this point, then revelation would soon be possible.

Of course it was not. Another major concern was that of preventing the Russians from learning how much uranium the United States was obtaining. It was for this reason that Marshall directed that uranium not be listed by Belgium as one of her resources as she met with other countries to compose the European response to the offer of Marshall Plan aid. This was fair, it was also argued, because industrial application of atomic energy was so remote.[61] The annoyance Kirk expressed in late summer over an article printed in the Paris edition of the *New York Herald Tribune* on uranium shipments and income received was because "the Russians can now deduce pretty closely how much active stuff we got from the Congo deliveries" if they did not already know.[62] Apparently some of Godding's instructions had been overlooked and subordinates had continued to compile and publish reports which were meant to be kept secret.

Sengier's fear of war with Russia by now was such that he surreptitiously arranged for all his papers to be shipped to New York and to be granted full powers for directing the company, as he had during World War II. No one in the Belgian government knew of these moves. Should a crisis develop, it would be up to the Americans to fly him out of Belgium or the Congo. During the last week in May while in Paris, Sengier had been trailed and saw a Russian take the room next to him in his hotel. Kirk assured the Belgian that he would be taken care of and reminded Acheson that "his physical presence [in] America in event of war is indispensable to our receipt [of] ores from [the] Congo."[63]

Thus when Spaak and Sengier both planned visits to New York in September 1947, the time was ripe for in-depth discussions. Spaak was arriving to serve as chairman of the Belgian delegation to the UN and Sengier to negotiate a new contract. The talks would be all the more important as the United States saw itself faced with a critical shortage of natural uranium, with the only major source of supply the Congo.[64]

Invitation of Belgian Scientists to the United States

Acheson had by now resigned as under secretary and been replaced by Robert Lovett. Kirk took pains to brief his new superior. He pointed out that Sengier thought the two governments were getting the ore "dirt cheap." The chief recompense to Belgium would be under Article 9(a). Sengier would also be informing the AEC of his tax difficulties. Though Kirk did not say as much, the possibility that Sengier would not be satisfied with a continuation of agreed-upon prices was implied. Spaak had never indicated he thought the contract terms were unfavorable to Belgium, but he might be forced to yield to internal pressures in order to stay in power. Kirk felt sure that Spaak, Sengier, and the prince regent would hold firm against Communist maneuvers, which currently seemed to be focusing on the issue of price. But he had an "uneasy suspicion" that Wigny was hostile to Sengier as a big businessman and might be contemplating further action in the Congo, such as nationalization.[65]

In order "to meet M. Wigny a little if he gets too active," Kirk suggested that Lovett review any prospective loans that Belgium or the Congo might try to obtain through the Export-Import and International Banks.[66] This recommendation, so similar to that proposed by Harvey Bundy when Sengier had been slow to sign in September 1944, was reviewed in general terms in a joint meeting of the secretaries of state, war, and navy and several of their staffs on 11 September. George F. Kennan, then director of the policy planning staff in the state department, argued that the projected aid program for Europe should stand or fall on its own merit and not be tied to the atomic materials issue. The United States would be subject to severe criticism were it known it was "bargaining relief aid for rights to atomic material."[67] Kennan was supported in his views by Secretary of War Kenneth C. Royall, and the matter was dropped.

Senator Arthur Vandenberg, chairman of the Senate Foreign Relations Committee, raised the matter in different form some weeks later. He suggested that the United States might pay for deliveries of strategic material only if repayment of long-term U.S. loans were kept up.[68] The state department continued to hold that uranium should not be made a *quid pro quo* for European aid. On 26 November Lovett reported to a high-level gathering that "any attempt to gain further concessions from the Belgians through European Economic Recovery legislation would appear as an attempt to obtain concessions already secured to us by committments entered into in mutual good faith."[69]

From the Belgian viewpoint at the time and in the perspective afforded by the years which have passed since 1947, the Belgian requests for openness and information sharing were reasonable. Yet it should be remembered that in the fall of 1947 no Westerner knew for sure where Soviet research on atomic devices stood; any leak of information could be disastrous. Nor did the United States and United Kingdom have much information applicable to industrial developments that could safely be divulged. The Belgians' sense that they were being put off, however, would only be accentuated by the report that Britain was building an atomic energy plant. So eager were the scientists of the small country to be active that they informally proposed building an atomic pile alongside that of the British, to send electricity by cable across the Channel. Locating their project in England would avoid the issue of Belgium's vulnerability to invasion from the east.[70] If the two English-speaking countries did not wish to accede to Belgian requests and were to forego heavy-handed dollar diplomacy, which could backfire seriously, then they had to rely on the good will and understanding of Spaak and Sengier. The tenuousness of the entire situation was underscored by Ambassador Kirk's recommendation that Lovett see the two men separately since they were not politically close. How long could the team of Spaak and Sengier be held together?

When Spaak met with Marshall in New York on 3 October 1947, he made clear that the accord was causing increasing difficulty, as now even non-Communists were making inquiries. Yet he himself did not feel at liberty even to inform his fellow cabinet members. He had no objection to making the agreement public but would not press that point if he could at least say that Belgian interests were safeguarded by provisions for his country to share equitably in the benefits of industrial utilization. This was his basic point, and he asked about the possibility of Belgium receiving electricity from a second atomic pile in England. Marshall heard the premier out, promised a response, pointed out the danger of what a skilled propagandist could do with any public statement, and admitted his unhappiness over the construction of even one pile in the United Kingdom.[71]

The British and Americans were, in fact, having considerable difficulty sharing information and uranium between themselves. Until these problems were worked through, it was unlikely that serious talks about implementing the exchange of industrial atomic information foreseen by Article 9(a) of the Tripartite Agreement could begin with the Belgians. Meanwhile, Spaak dealt with questions in the chamber in the same sort of phraseology as in the past. Sengier, who was not aware that Spaak was speaking in conformity with the state department's wishes, thought the prime minister might have done better with a more frank and complete statement. As far as industrial use of uranium was concerned, Lovett gave Spaak the text of a speech to be given by David Lilienthal, chairman of the AEC, which made clear how remote such use was and how unjustified foreign hopes were. American scientists also told Spaak that the idea of piping electricity across the Channel was a fantasy.[72]

Sengier apparently was satisfied with his own meetings with Lovett and especially the AEC. They reached agreement on a new contract for still more ore at a higher price, which presumably brightened the profit position of the Union Minière, somewhat lightening the impact of Wigny's tax.[73] The United States greatly needed more uranium to sustain its current nuclear program. The joint chiefs of staff and the Atomic Energy Commission saw any reduction as having an adverse impact on national security and argued that security required expansion of the energy program and "preemption of as much as possible of world production of uranium."[74] In these circumstances, the AEC was no doubt inclined to be generous in its contract negotiations.

Though given some information quickly via the Lilienthal speech and American scientists, Spaak got no prompt response on his basic request for implementation of Article 9(a). Just before Christmas he again asked how he could justify the construction of an atomic pile in England fueled by Belgian uranium unless Belgium participated on equal terms. The prime minister complained that he was being placed in an untenable position. On 19 January 1948, Belgian Ambassador Silvercruys reminded the state department of Spaak's desire for a response to his demarche of the preceding October.[75]

By this time a *modus vivendi* had been worked out among the British, Americans, and Canadians regarding the sharing of information and distribution of uranium supplies. Marshall therefore hoped to get his response to Spaak by the middle of February, but the drafting took longer than expected. Meanwhile a minor incident again demonstrated to the Americans how much they were indebted to the Belgian premier.

What had happened was that, while in Moscow, a Belgian trade delegation had agreed to a Russian request that it present to the Belgian ministry of foreign affairs a draft letter from the Belgian ambassador in Moscow to the Soviet foreign office. The letter stated that while Belgium could not supply any uranium to the U.S.S.R. in 1948, it would examine the possibility for 1949. Spaak recognized that the delegation had been outwitted by Soviet Minister of Foreign Trade Anastas Mikoyan and had acted precipitately and in excess of its authority. The prime minister immediately informed the Soviets that during the war Congo uranium had been placed at the disposal of the United States and the United Kingdom and that there was no question of examining the matter of supplying Russia in 1949.[76]

It was a close call. Because of official secrecy, Belgian representatives knew less about their own country's dealings than the Russians probably did. Yet when Marshall finally replied to Spaak's 3 October request more than five months later, he stated that while many details of the agreement might be known, he still did not believe full revelation would be prudent in view of increasing world tensions.

A disclosure at this time would at the least stimulate speculation as to amount and tempo of individual ore shipments, our degree of dependence on the Congo, and the relation of the Congo to the over-all procurement program. Out of such speculation some details might be deduced about our bomb production rates. It is just possible that some of the gaps in the Soviet Union's estimate of our position might be filled.[77]

Moreover, disclosure would enable the U.S.S.R. to claim that the American motive in backing the European Economic Cooperation plan was insurance of supplies of uranium. Finally, Marshall did not think it wise to release anything until the United Nations had finished its discussions of the matter of atomic energy control.

As for Article 9(a), Marshall made the usual promises, followed by the disclaimer that the primary American research emphasis was still strategic and that "tremendous technical difficulties" still had to be overcome before industrial use could be attempted on anything other than a token basis.

Then came the small concession worked out in the preceding weeks, no doubt under prodding from Kirk, to the effect that at least some demonstration of professional good will should be made. Progress had been made in medical and research use of radioisotopes. Marshall suggested representative and technical personnel from the two countries could develop means of sharing this information; in the process, further explanations could be made of the difficulties associated with commercial usage.

Even to this minor gesture the Americans felt required to add a special caution. They knew some members of the Communist party held executive positions in the Belgian atomic research program. This, Marshall said, would make it harder to convince his own people that it was in the national interest to share any information at all. Therefore a careful run-down on each participating individual would be required. The concession was limited and such as not to disturb nervous patriotic watchdogs. It was made, as the American members of the Combined Policy Committee put it, because "Mr. Spaak needs something of the kind to strengthen his hand vis-à-vis both the nationalistic and Communist elements in Belgium."[78]

The Belgian premier was quick to take advantage of the opening. When he accompanied the prince regent on a formal visit to the United States in the first weeks of April 1948, he had his specialist on atomic energy matters, Minister for Economic Coordination and Re-equipment Paul de Groote, visit with Gullion. A scientific mission to Washington was scheduled for August, and at the beginning of July the Belgians proposed its personnel: Professor Van den Dungen, the pro-rector of the Free University of Brussels; Professor Alexandre de Hemptienne, a physics and mathematics specialist from the University of Louvain; Professor Gueben, who led a similar position at Liège; and Mr. Paul Gerard, a mining engineer and consultant in De Groote's ministry.[79]

Kirk was shocked by the proposal of Van den Dungen, a man of leftist views and still personally close to his former students Libois and Geheniau, both declared Communists. In a recent conversation the secretary of the Administrative and Scientific Committee of the Inter-University Institute of Nuclear Physics, a man Kirk considered left-leaning himself, had described Van den Dungen as an idealistic Communist and a loyal patriotic Belgian who would side with Russia if that country were at war with the United States. The state department, also upset, suggested that the mission be limited to three men, and indicated the work of the group would be more constructive without Van den Dungen. Kirk negotiated with Spaak, who found a way out from the pressure of taking representatives of too many universities and viewpoints by naming only Hemptienne and Gerard, who both held national positions by virtue of their membership in the Inter-University Institute of Nuclear Physics.

The Belgians also prepared a seven-part program for discussion, but the Americans quickly cited the McMahon Act as preventing consideration of means of producing atomic energy without the intermediary use of fissionable materials, costs of various size units, or technical problems such as cooling, heat exchange, and radiation protection. Other items such as price costs compared with other forms of energy, certain metallurgical issues, and future research collaboration might be discussed, as well as the usual issue of when industrial atomic energy could be touched upon. The Belgian list included no mention of medical isotopes, the one area on which the Americans had indicated some openness.[80]

When the two Belgian scientists arrived in late August, Marshall personally told them that the time had not yet arrived when practical effect could be given to Article 9(a). Though this may have been a disappointment to the scientists, three weeks later Lovett thought the men left the country happy with their experience. They had been discreet in their questions and conduct. As a result, the United States was prepared to enroll qualified Belgian students in the unclassified radioisotope school at Oak Ridge, to grant more extensive use of isotopes, to enroll more Belgian students in American colleges to work on physics and chemistry with no direct link with the AEC save a check on the students' political dependability, and to expedite requests for equipment. The state department also gave the professors a sheaf of unclassified documents and promised to recommend to the AEC shipment of a requested ten liters of heavy water as an amount so small as to be exportable.[81]

The publicity attending the visit of the Belgian scientists assuaged Belgian pride and seemed to ease some of the political pressure on the issue. The excesses of Soviet policy in the spring of 1948 in Berlin, Czechoslovakia, and elsewhere so discredited Communists in Belgium that Spaak's position was strengthened and their badgering reduced. The promises of technical assistance which Carroll L. Wilson, general manager of the Atomic Energy Commission, extended to the Belgians also facilitated the negotiation of a new contract for 5,000 more tons of uranium ore that December.[82] Yet what had been achieved was small. Belgium had made an enormous and invaluable contribution to the war effort and subsequent security of the West. In return, a good price had been received, plus the promise of future inclusion in the industrial atomic energy club which might significantly aid Belgium's economy and future industrial profits.

On the other hand, limited as the arrangement was and brief as the scientists' visit might be, these nevertheless represented the emergence of a major issue: proliferation of nuclear capabilities, not through participation in the original scientific work, by spying, or by independent discovery, but by diplomacy. Inherent in points raised by Gutt in 1944, its embryo was evident in Article 9 of the Belgian accord. In the context of the war, little thought was given to the view that small powers might take toward the monopoly of atomic energy by a condominium of a few great powers; the growing confrontation with Russia overshadowed such concerns, and the rise of the Third World was only imperfectly forecast. By the end of 1947 the implications of the promises made to obtain uranium were becoming more clear.The greatest commitment was to Belgium, but surely the demands for information would spread to the other supply countries.

The secrecy and lack of implementation of Article 9(a) insisted upon by the United Kingdom and the United States prevented Belgian governments from reaping the prestige initially envisioned, and the very size of the profits turned that issue into a potential political liability for them. Decision makers within the two English-speaking governments seemed surprisingly insensitive to the importance of matters of pride and prestige for a small country still reeling from defeat, occupation, economic misery, calumny for alleged too rapid surrender to Nazi forces in 1940, and which was fiercely divided by linguistic issues and even over continuing the reign of its king. Surely a central factor governing attitudes in Washington and London was the persistent fear of Russia, Communists and Socialists in general, and of security leaks which might result from information sharing. The last issue, together with differences with the British over uranium allocation, accounted in substantial part for the delay in Marshall's response to Spaak's inquiry, and it is to these matters that attention must now be turned.

5 · Reluctant Anglo-American Collaboration

Successful completion of agreements which assured the United Kingdom and the United States purchase or control of the chief known sources of uranium and thorium in the free world did not mean that supply problems were at an end. United States officials constantly feared that all shipments combined would be insufficient for American research and production needs. The British and Canadians wished to speed their own programs; other countries, especially Belgium and France, also envisioned development of nuclear research programs. Yet such proliferation meant diffusion of scarce materials, pressure for exchange of information, increased danger of security leaks, and possibly the appearance of new possessors of atomic weapons.

The decision of American and British statesmen to try to achieve some international control of nuclear energy through the United Nations and the frustration of their attempt is a long, complicated, and somewhat separate story from the effort to acquire a monopoly of uranium supplies. The two policies at times seemed almost contradictory, yet it was natural that the Americans and British, while attempting to create an arrangement for international control, did all they could to assure their own security. More directly related to the search for raw materials were the controversies that shook the Combined Development Trust, as proposals for division of the materials were discussed and the American congress clamped down on information sharing.

Early Contacts and the Quebec Accord of 1943

The possibility of scientific exchange on war-related research between Britain and the United States had been raised on British initiative in July 1940; by April 1941 appropriate missions were in place. Uranium and atomic research were not central items of discussion. When American feelers were extended regarding closer collaboration on these matters, the response from the British was starchy and cool. Unaware of the recent surge of the American effort, they believed they were far ahead in research and thus had little to gain; moreover, they considered U.S. security insufficient. Eventually a successful visit to North America by British scientists made them aware of American progress and the value of information exchange.

Discussions were held, and it was agreed that when the time came for a production plant to be constructed, it should be located in Canada. In conversations between President Roosevelt and British Prime Minister Winston Churchill at Roosevelt's home in Hyde Park in June of 1942, oral agreement was also reached that the two nations should work on atomic projects on completely equal terms. Because of the impracticality of erecting research plants within the range of German air reconnaissance and possible bombing in England, Roosevelt said that the United States would do the job; British resources should not suffer such a major diversion from more immediate war efforts. Though little was said publicly, it now appears that the Americans felt insecure about production of such a weapon so close to the continent and wanted it moved across the Atlantic.[1]

The absence of any written formula on information sharing and the concentration of production in the United States had their natural consequences. As U.S. efforts leaped ahead with the assignment of high priority and the appointment of Groves as ramrod, the Americans came to realize that they were doing far more than the British. Their view of a nine-to-one imbalance of effort and investment, together with disagreements about military strategy, led to an evasive response to British attempts to increase information exchange on production of fissionable materials.

The argument for information sharing was basically that of enhancing the war effort, of winning earlier. The argument against was the need to preserve secrecy. Since the great proportion of production work was being done in the United States, the amount of aid that British scientists and engineers could give would be limited; therefore why risk additional security leaks, some American officials grumbled. Only restricted exchange on basic research, not on manufacture and production, should be permitted. This they told Roosevelt when it became obvious that a more specific, written policy was needed instead of memory of vague and friendly exchanges at Hyde Park. A report at the end of September 1942 of an unexpected Anglo-Russian agreement for exchange of information on new and future weapons further helped to nudge the president to support a limited exchange policy. The British, piqued by the American behavior and confident of their own work, rapidly became less interested in sharing.

Their annoyance was increased manyfold by a January 1943 memorandum or unofficial working paper issued by Dr. James Conant as chairman of the American National Defense Research Committee, which sharply circumscribed, on the basis of "need to know," information to be shared by the Americans with the British and Canadians. Anderson in Britain retaliated by ordering that contacts be reduced until full reciprocity had been restored; this, he thought, might strengthen the prime minister's bargaining position.[2] By early spring of 1943 interchange on nuclear weapon research and production had nearly halted.

When Churchill learned of the impasse, he pressed for reinstitution of full information sharing. U.S. opposition to such a move in the levels beneath the president was at the same time hardening. For one matter, the principle that information should be exchanged only for use-in-this-war had been generally accepted. Yet one of Churchill's chief science advisors, Lord Cherwell, admitted that while the British could not hope to produce a bomb during the war, they planned to do so promptly after its conclusion in order to assure their safety. A second matter was strong American suspicion that one or more of the key British science representatives, such as W. A. Akers or even Sir John Anderson, wanted the information less for bomb production than to assist Imperial Chemical Industries in developing nuclear power plants which would enable that firm to reap immense commercial profits in the postwar years.

British interest in exchange, on the other hand, was quickening. A careful examination of the costs of erecting British plants for nuclear research and production showed London officials in January 1943 how terribly exhausting the effort would be. Much of the nearly prohibitive cost would be spent in duplicating American activities already underway. At the beginning of May came the discovery that Groves had managed to buy up not only the entire Eldorado uranium output for the next few years, but also all the heavy water—an item important for nuclear research and production of a certain type of atomic pile— manufactured by a major Canadian plant. If the British were to go it alone, they would be doing so without sufficient supplies of necessary materials. Anderson telegraphed Churchill, who was on his way again to meet Roosevelt, that the need for collaboration was urgent.[3]

Roosevelt, uninformed of the concerns of his lieutenants, permitted himself another sweeping yet vague promise of cooperation in response to Churchill's nudging in the May 1943 talks. In ensuing weeks Churchill pressed relentlessly, and Roosevelt could not retract. By now strategic issues required the Americans to avoid minor irritations in the alliance. Many American scientists and military men, as well as some Canadians, were nevertheless annoyed by the way the two leaders had pushed ahead. Vannevar Bush, director of the Office of Scientific Research and Development and one of the scientists better in tune with administration views and problems, therefore took up the matter with Churchill. The prime minister disavowed any interest in postwar matters and promised that Roosevelt could limit British commercial uses in such manner as he thought fair, given the greater American investment.

A breakthrough was achieved. In the following weeks, Sir John Anderson forwarded proposals to Bush which called for a joint effort, stated that the powers would never use the weapon against each other and would require mutual consent before using it against a third party, and provided for the creation of a Combined Policy Committee (CPC). This committee was to allocate the tasks undertaken by each country and the distribution of scarce materials, as well as generally to interpret and apply the agreement. A hasty visit by Anderson to Washington to confer with Bush, Conant, and others brought a meeting of minds. The proposals would form the heart of the "Articles of Agreement Governing Collaboration between the Authorities of the U.S.A. and the U.K. in the Matter of Tube Alloys," signed by Roosevelt and Churchill 19 August 1943 at their meeting in Quebec.

Though at Quebec emphasis was placed on collaboration in the current war, Anderson made clear that he was motivated too by concern that Britain not have to "rely entirely on America after the war should Russia or some other power develop" the bomb.[4] This claim for nuclear independence was conveniently ignored and was to cause much trouble when it reappeared with vigor in 1946. For the time being, an end had been brought to a too-long period of misunderstanding, caused in part by overstrong egos unaware or unwilling to admit the sacrifices and achievements made by either side and acerbated by an initially unwise choice of emissaries by the British. The political approach by Churchill simply annoyed American scientists, and the use of W. Akers and M. W. Perrin, former high-level employees of Imperial Chemical

Industries, provoked suspicions that eventually forced the recall of Akers. The broad overview from the political top was, however, what was needed to counteract the more parochial views of the lower echelons, and it was the personalities of Bush and Stimson, as well as of Dr. James Chadwick (discoverer of the neutron and Akers' replacement as head of the British scientific mission in North America), which did much to overcome the impasse. If poor choice of personnel helped to cause the differences, later good selection of men helped to reduce them.

Growing Differences

At their first meeting on 8 September 1943, the three American, two British, and one Canadian members of the Combined Policy Committee took up procedures for interchange of information and scientists. Groves disliked the idea. Although he saw its usefulness in such areas as gaseous diffusion, where the British had worked for some while, for the most part he was not helpful and at the best was "carefully and unenthusiastically correct."[5] Nevertheless by the turn of the year British scientists were arriving on the scene, among them a team of experts on diffusion that included the emigré German theoretician Klaus Fuchs. The Canadian and British position is revealed by a diary entry made by Dr. C. J. Mackenzie, head of the Canadian Research Council:

> It is apparent that Chadwick feels, as I have always felt, that the United States' effort is one hundred times greater than any possible United Kingdom effort, that the Americans can get along if necessary without the U.K., while the U.K. can do nothing without the U.S.; and that our best policy is to make sure we get collaboration and also that we will have to change our plans to suit the American situation.[6]

Bush and Anderson were pleased at the resumption of collaboration. So too were Churchill and Roosevelt. Meeting again at Hyde Park in September 1944 following the more formal Second Quebec Conference, the two signed an important *aide-mémoire*. It stated their view that work on the bomb should continue in utmost secrecy, rather than with publicity which might initiate international control, as Danish physicist Niels Bohr was advocating. Use of the weapon against Japan was envisioned, as was continuing full collaboration for commercial as well as military purposes following the fall of Japan.

Several individuals in the American program were wary of such a commitment on postwar cooperation; they had avoided broaching the topic until they had worked out their own views regarding the wisdom of international control as compared with bilateral or single-handed research. Yet in the flush of friendship associated with the growing military successes, the busy and tired Roosevelt made a commitment without extensive prior consultation with his aides. Indeed, he talked about the issues with Bush, without telling him of the *aide-mémoire*, only several days later. Bush was more than a little disturbed, for he feared Anglo-American secrecy "might well lead to extraordinary efforts by Russia to develop the bomb secretly and to a catastrophic conflict, say twenty years hence."[7]

Even as the debate about policy was growing, an incident involving French scientists sharpened American doubts about collaboration with the British. The procession of events was complex, and eventually a great many memoranda were generated, tracing the developments and debating the issues. During the war, some French scientists who in 1940 escaped Nazi troops had participated in British research undertaken at a plant in Montreal, Canada. Dr. Hans van Halban had even brought with him much of Joliot-Curie's valuable heavy water. After careful negotiation with the Frenchmen, the British had by August 1942 struck a deal. Britain would receive the patent rights held by two of the French scientists. In return, Britain promised to reassign to France rights to those patents for French territories and similar rights in any *future* patents that the visiting French scientists might obtain.

The Quebec Agreement of 1943 stated that neither participant would reveal information from its nuclear research to a third party. Yet at that time the British had failed to inform the Americans of any former and conflicting obligation. Now the French scientists wished to return to their liberated homeland, some obviously anxious to confer with their former mentor Joliot-Curie. The British supported their request and facilitated their travel. Groves was furious, for the loss of secrecy was obvious. He protested, and eventually a compromise was crafted which delayed the travel of some and curtailed other contacts. The incident confirmed, however, the determination of Roosevelt and Churchill to hold to secrecy and to avoid any discussion of the matter with the French, whom Churchill in particular suspected might funnel information to Russia as a counter to the influence of Britain and the United States in Europe.[8]

As the allies began the year which was to bring military victory, there was disagreement in the American camp on policy regarding continuing secret collaboration with Britain, in both theory and practice. The official historians of the U.S. Atomic Energy Commision have correctly observed that episodes such as that of the French scientists "added tension to the mistrust and misunderstanding that had accumulated through the years. It made it difficult for the wartime partners to face the future in wisdom as well as strength."[9] Yet if the Americans were slow in collaborating, out of desire to reduce competition with the Soviets until international control was achieved, to protect against security leaks, and to avoid giving British industry any postwar aid, they would be violating the terms of the Hyde Park *aide-mémoire*, the existence of which Bush and Secretary of War Stimson were still unaware weeks after Roosevelt's death in April 1945. The British reaction to the slowness of implementation could only be dismay.

The French scientist imbroglio reflected in addition to British-American differences some significant Anglo-Canadian problems. Halban, the brilliant French emigré and former disciple of Joliot who was sent from Britain to take charge of the atom research team at Montreal, was not a good administrator. He soon came in conflict with Mackenzie of the Canadian Research Council. Mackenzie found Halban arrogant, while Anderson in London came to dislike Mackenzie for not wholeheartedly backing every British viewpoint.

The Canadians' position required that they work well with their giant American neighbor. They themselves questioned the international make-up of the Montreal team and shared American concerns about security. Conant's January 1943 restrictions were not entirely unreasonable, they thought; at least some progress could be made under them and Canadian research could move forward. They resisted Anderson's orders to cut off contacts with Americans. Like other small and medium powers desirous of maintaining their own dignity and independence, Canada endeavored to hold a middle position, attempting to mediate between her two great friends rather than walking in lock step with British policy, as Sir John desired. In the French scientist affair, Canadian Minister of Munitions C. D. Howe complained bitterly that he had been placed without warning in the middle of an unpleasant situation which endangered American-Canadian working relations and progress of Canadian research. Given Howe's failure to prevent sale of Eldorado uranium to the United States, Anderson was likely not sympathetic.[10]

Anglo-American differences over information exchange were bound to affect the uranium acquisition program. They did so only slowly, however. This was in part because it was understood that all uranium would be assigned to the United States for weapons development until the Japanese surrender. Moreover, Groves was aware that some rare materials, such as those in Travancore and South Africa, could be obtained only with the aid of the British. He therefore strove to avoid friction in the procurement program, although his desire not to be too dependent on the British was evident in his cultivation of Sengier and in the Swedish negotiations.

At a meeting of the Combined Policy Committee in Washington on 4 July 1945, Harvey Bundy, one of the committee's secretaries and an assistant to Stimson, suggested that the committee once again go on record as endorsing the allocation of all shipments of ore to the U.S. weapons program. British Ambassador Lord Halifax, who attended on invitation, endorsed the arrangement for the duration of the war but pointed out that at war's end it would leave the United States with atomic weapons and Britain without even any raw material. It was then agreed that any materials received beyond those needed by the United States for weapons in the current war should be held by the Combined Development Trust for future allocation.[11] Further warning of future difficulties on allocation of scarce ores was implicit in the announcement by Halifax in October that the British government planned to set up a research establishment on atomic energy, which effort would include creation of an atomic pile.

Meanwhile negotiations had moved to relatively satisfactory conclusion with Brazil, the Netherlands, and Sweden. Within the leadership of the United States, debate continued between those, such as Bush and Stimson, who now favored a policy of divulging information in connection with creation of a system of international control, and those such as Groves and Secretary of the Navy James V. Forrestal, who favored continuing secrecy. On 21 September, Stimson retired on his seventy-eighth birthday and was succeeded by Judge Robert P. Patterson, his former under secretary. Patterson had been won to Stimson's views, but found that Secretary of State James F. Byrnes was more than a little dubious about the possibility of working with the Russians.

The November 1945 Agreements

In London, Labor Prime Minister Clement Attlee, victor in the elections of July 1945, was facing sharp questions about the atom. Though he and his foreign minister, Ernest Bevin, were initially attracted by the notion of information sharing with the Soviets as a means of maintaining good relations, they soon took a tougher line. Attlee pressed Truman for a meeting with himself and Canadian Prime Minister W. L. Mackenzie King on the subject. Despite Byrnes' efforts to procrastinate, it was necessary to schedule a conference for the middle of November. There was need for joint agreement on an international approach to control of the atom and also on how further interchange between the two powers was to be conducted. Concurrence on this was all the more pressing because the old Quebec Agreement no longer seemed satisfactory. As for the Hyde Park *aide-mémoire,* the Americans did not know of it and had to receive a copy from the British. What interchange still existed at war's end had narrowed to virtually nothing, yet much time and expense could be spared the British, who now were starting their research station at Harwell.

Preparation for the British visit was complicated by strong currents seething through Washington over the atomic energy control bill which the young democratic senator from Connecticut, Brien McMahon, was planning to introduce. Many of the scientists who had labored for the Manhattan District Project protested the restrictions that the bill appeared to impose on their future research. The greatest furor centered on whether the military should be excluded from future control of atomic energy. Should they be banned not only from seats on the proposed Atomic Energy Commission but also from administrative and staff posts? The civilian-military issue became a major point of contention, fueled by strong personality conflicts between McMahon and his supporters and Groves. Little attention was paid to a section of McMahon's bill which stated that basic scientific information could be freely disseminated but should be distinguished from processes or techniques. Related technical information could be released by a special board, but only if "not of value to the national defense" nor to persons considered ineligible under the Espionage Act.[12]

It had been understood for a number of months that the theories necessary for production of atomic weapons were so widely known that there was no sense to concealing them. It was the engineering features—the controlling of fission, the construction of diffusion plants—that had proved so difficult and was achieved in such secrecy by the Manhattan Engineering District. This information was central to the construction of atomic piles suitable for either military or commercial purposes, and this was the sort of information the British, Belgians, and others desired. The importance of this distinction had been well discussed. Its implications for existing international agreements, however, was not fully canvassed because the terms of those wartime agreements were not widely known. Indeed, the U.S. congress was not even aware of the existence of the Combined Policy Committee or the Combined Development Trust, both the creatures, direct or indirect, of the Quebec Agreement of 1943; nor did the legislators know of the Tripartite Agreement of 1944 and its Article 9.

At the beginning of November Patterson offered Byrnes the assistance of the war department in preparation for the arrival of the British. Byrnes was noncommittal, but Patterson forged ahead by asking Lieutenant R. Gordon Arneson, who had been working closely with Groves, to develop a study of existing obligations based on Bush's previously expressed views. This was promptly submitted, along with proposals for consideration with the British. These were revised by a group including Groves, Bush, and George L. Harrison. The latter was the alternate chairman of a special Interim Committee established in May, under the nominal chairmanship of the secretary of war and consisting of knowledgeable civilians, to make recommendations regarding postwar research and control.

These men endorsed a cautious approach toward international control, including some feelers toward Russia. They recognized that the British wished clarification or release from Clause IV of the Quebec Agreement, which stated:

> The Prime Minister expressly disclaims any interest in the industrial and commercial aspects beyond what may be considered by the President of the United States to be fair and just and in harmony with the economic welfare of the world.[13]

The American position was simple and no doubt accorded with Groves' desire to get control of as much uranium as possible. Clause IV would be abrogated in any new arrangement, in return for a promise that the United Kingdom bring all uranium and thorium ores in the Commonwealth under control of the Trust and subject to allocation by the Combined Policy Committee. The British would still have a voice in allocation, but this arrangement would give the United States an equal voice on materials over which it otherwise would have no control whatsoever. Groves, having replaced Bundy as secretary of the CPC, was sure to have his views heard. Bush by this time had reached the position that while he could support sharing raw materials under a revised Quebec Agreement, sharing of technical information should await a broader international agreement.[14]

While Truman, Attlee, and Mackenzie King dealt with the large issue of international control, Patterson and his group met with Sir John Anderson's team. The latter's presence was something of an anomaly. An independent member of parliament for the Scottish universities, he sat on the opposition front bench though not a member of the Conservative party. He nevertheless had been asked by the Labor prime minister in September 1945 to chair an advisory committee on the nation's program in nuclear physics. Few men within the Labor government knew anything about the issue—even Attlee had been kept in the dark by Churchill despite his position as vice prime minister—and so Anderson's expertise and familiarity with detail was needed. Anderson, however, did not have true ministerial status, and in time his committee would be undercut by an Official Committee on Atomic Energy formed in August 1947. Then, too, in Britain other small *ad hoc* or "Gen" committees grew up in profusion succeeding the initial Gen 75 committee of some seven cabinet members that held forth from August 1945 to December 1946. By December 1947 Anderson's presence was no longer so essential. Moreover, while cooperative on atomic energy matters, he was attacking the Labor government's other policies sharply. At the end of 1947 he would resign his chairmanship, and his committee would be disbanded. As the official historian of the United Kingdom Atomic Energy Authority has written, the committee system of the U.K.'s effort was indeed "labyrinthine."[15]

When Anderson inquired about Clause IV at the Washington meeting in November 1945, U.S. Secretary of War Patterson indicated willingness to arrange a solution which "would not place the U.K. at a disadvantage."[16] All agreed that the Quebec Agreement should be replaced *in toto*. Anderson thought any new accord should involve full interchange of personnel. Groves quickly suggested that the *quid pro quo* for this might be British transfer of all uranium and thorium to the control of the Combined Development Trust for allocation. Minutes of the meeting show that Anderson knew at what Groves was aiming:

> In agreeing with this point, Sir John pointed out that the U.K. would have to proceed with caution in some cases, as for example, South Africa. General Groves expressed the view that South Africa would probably agree to sell its ore to the Trust. Sir John agreed that since South Africa had no establishments built she would probably be willing to sell and that the U.K. or the U.K. and U.S. jointly might approach her soon.[17]

It was agreed that the Combined Policy Committee and the Trust should continue, and a subcommittee was charged with formulating a Memorandum of Intention spelling out the policies to be followed in rewriting the Quebec Agreement.

In the sub-group, Groves and Harrison argued for two memoranda, one a short statement from the heads of government directing the CPC to prepare a new agreement for continuation of that committee and the CDT. The other, longer note would summarize the policies to be followed in the new agreement. The British went along, but soon a divergence of views arose. Groves wished a restrictive formula which would allow cooperation and exchange in basic research but would leave interchange on design, development, construction, and operation of large plants to *ad hoc* arrangements through the CPC.

This displeased the British. The phrasing stayed, but in full committee Anderson insisted that the first, short memorandum refer to full and effective rather than just effective cooperation. Groves bristled, but Patterson acquiesced. The secretary of war also accepted Anderson's contention that the long memorandum was only a general guide rather than a binding set of policies, again a defeat for Groves.[18]

Anderson had done well. In return for granting joint control of Commonwealth materials, he gained not only omission of any reference to commercial rights but also the possibility of reestablishing the sort of exchange originally envisioned at Quebec and which had been whittled away over the years. The CPC was to be maintained, and Canada was to be admitted as a full partner. Only prior consultation would be required, rather than mutual consent, for use of the bomb against third parties. Groves did obtain leverage to restrict the flow of technical information if he could convince the CPC to follow his views. Important for the uranium hunt was agreement to "use every endeavor with respect to the remaining territories of the British Commonwealth, and other countries to acquire all available supplies of uranium and thorium."[19] These supplies were to be allocated to the three governments; any left over beyond their needs were to be retained by the Trust for later dispersal as conditions warranted.

The surface harmony expressed in the signed documents only briefly concealed discord about fundamentals. Some Americans feared too close collaboration with the British might prevent, because of Russian suspicions, the erection of a multinational system of control of the atom. Representatives of the American military, especially Groves, forcefully argued that, for security reasons, the sharing of information should be more limited than the British desired.

The task of drafting a formal agreement along the lines of the longer November Memorandum of Intention was referred by the CPC to a subcommittee of Groves, Lester Pearson of Canada, and Roger Makins of the United Kingdom. At the same time, the American government was formulating its proposals for more general interchange of scientists and scientific data. These envisioned exchange of radioactive isotopes and visits to basic research laboratories. Secretary of the Navy Forrestal was quick to counsel against too rapid disclosure, and Groves warned against any discussion of raw materials.[20]

All this occurred at the same time controversy was mounting over exclusion of the military from the proposed American Atomic Energy Commission. The hostility between Groves and McMahon was no secret. On 2 January 1946 Groves drafted a memorandum which reflected his frame of mind.

> [T]he United States must for all time maintain absolute supremacy in atomic weapons. . . . If we were truly realistic instead of idealistic, as we appear to be, we would not permit any foreign power with which we are not firmly allied, and in which we do not have absolute confidence to possess atomic weapons. If such a country started to make atomic weapons we would destroy its capacity. . . . Either we must have a hard-boiled, realistic, enforceable, world-agreement ensuring the outlawing of atomic weapons or we and our dependable allies must have an exclusive supremacy in the field, which means that no other nation can be permitted to have atomic weapons. . . . Our military establishment must not be excluded from research and development in the atomic weapons

field. . . . If there are to be atomic weapons in the world, we must have the best, the biggest and the most; and the Army and Navy must not be divorced from their responsibility of defending the United States.[21]

Groves' annoyance about what he considered the murkiness of U.S. policy was linked to his realization that others were determined to curb his influence. He had been a tsar during the war years, accruing power and independence of action. Now the Atomic Energy Act debates required that his secrets be shared, and this stimulated open opposition to his power.

The uncomfortable duty of informing Under Secretary of State Dean Acheson, at the turn of the year, of the various commitments entered into by Groves and his colleagues fell to Joseph Volpe. Nearly forty years later, he still remembered how "absolutely flabbergasted" Acheson was and the under secretary's comment: "Why, you sons of bitches set up your own state department!"[22]

On 16 January 1946, according to David E. Lilienthal, director of the Tennessee Valley Authority and future chairman of the civilian Atomic Energy Commission (AEC), Acheson complained that "The War Department, and really one man in the War Department, General Groves, has, by the power of veto on the ground of 'military security,' really been determining and almost running foreign policy."[23] Negotiations on uranium with other states were not known to the state department, and Groves' ability to deny access to facts that would prove or disprove his theories about secrets made it impossible for any civilians to know if the country was or was not on the right track in its foreign policy. Acheson was determined that the situation not continue.

Already beleaguered by domestic legislators and now facing significant challenges within the Washington bureaucracy, Groves was outmaneuvered on the CPC's drafting committee, where the joint British-Canadian viewpoint prevailed and Groves' own stature as father of the Manhattan Project was negated by that of Pearson and Makins. London pushed for wording more specific than that agreed on in November, fearing that the provisions on raw material, where the British would make concessions, would be all too clear, while those for information, where the Americans were expected to yield, would be vague. Makins did the best he could but advised his superiors of the need for speed; the wartime forces for cohesion were rapidly weakening, and delay might spell loss of any agreement.

Groves was forced to go along with the mid-February proposal of the subcommittee. It followed the lines of the November Agreement. Regarding Clause IV of the Quebec accord, President Truman was simply to write Attlee that the United States government had decided it would be fair that there be no restrictions on the United Kingdom's development of commercial and industrial atomic energy. As for information exchange, there was to be "full and effective cooperation"; no distinction between basic research and applied technology was mentioned.[24]

Presentation of such a draft to the CPC did not imply Groves' support. Two days before its 15 February meeting, he wrote a long memorandum to Byrnes, who now served as chair of the CPC. Groves argued that the scope of the proposal "extends far beyond that ever contemplated by the Quebec Agreement" and in effect created an "outright alliance."[25] A few paragraphs later he repeated that "The proposed form of cooperation in the field of atomic energy could well be considered as tantamount to a military alliance." The November Agreements called for cooperation, not a joint venture. Moreover, did not the United Nations charter require registration of all international agreements and treaties? Word of a tripartite accord might undermine efforts for international control. The British were announcing intent to build a large-scale plant, and Dr. Chadwick had already indicated he would ask for "practically all of the processing techniques and plant designs and specifications of the entire Manhattan Project less the gas diffusion method, and even some of the personnel." Groves reported that he and Army Chief of Staff General Dwight D. Eisenhower believed that any cooperation should take into account that large-scale plants should be kept out of England; the British could build in Canada more safely. Byrnes was known not to be an Anglophile, and indeed considered in London as anti-British; Groves' arguments therefore no doubt received a favorable hearing.

The general also hastened to a man more inclined to the British, Under Secretary of State Acheson, who at this time was uninformed of the November negotiations. Groves pleaded for him and Byrnes to find some way to get the United States out of the "mess" into which Groves thought it was placed by the promise to give some technical assistance.[26]

Time was bought, as the question of registration of any agreement with the United Nations was referred by the CPC to legal experts. Soon it became evident that publicity had to be avoided so as to prevent Soviet suspicions and criticism. The nature of any continuing collaboration was therefore left in limbo. The delay, welcomed by the Americans, annoyed the British.

Compromise on Allocation of Materials

The pause in discussions persisted far longer than initially anticipated, for news broke the next day, 16 February 1946, which turned the tide of congressional debate on the McMahon bill and chilled consideration of further information exchange. Twenty-two persons had been arrested in Canada for disclosure of secret information, some to a foreign mission. Especially disturbing was later word that Alan Nunn May, a British scientist working in Canada who had visited a U.S. production installation in 1944, had passed information and material to the Soviets. Among the latter was a sample of U-233; this betrayed American interest in thorium, for only from it could U-233 be derived. Groves was highly disturbed.[27]

Repercussions of the discovery of the Gouzenko spy ring (so-named for its head) forced an amendment to McMahon's legislation which granted an important role to a Military Liaison Committee (MLC) that would work with—almost supervise—the Atomic Energy Commission. The importance of the atom to national security was such that it was now believed that those who were the chief guardians of the latter, i.e., the military, should not be excluded from a voice in atomic control. Groves was no longer viewed as a curmudgeon but as a hero.

Discovery of the spy ring strengthened Groves' hand on information exchange. When the British, indignant over the stalling, confronted Acheson on 15 March 1946, he told them it would be impossible to carry out the planned arrangement; to do so "would blow the Administration out of the water."[28] To evade the registration of an agreement with the UN by some exchange of notes would be fatal. The British would just have to accept that the United States could not fulfill the obligation and meet English wishes.

At the 15 April 1946 meeting of the CPC, Canadian legal authorities supported the American view on the registration with the UN of any new agreement. Lord Halifax, thoroughly annoyed, complained that the November Agreements were being left without effect and the United Kingdom without information needed for its atomic energy program. His effort to reword the old Quebec accord to fit the new circumstances failed. His disappointment was further embittered by the silence of Vannevar Bush, who Halifax hoped would support close collaboration but who in fact opposed it for fear it might embarrass the movement for international control.

This was not the only issue which upset the British ambassador. Prior to the February meeting of the CPC, its British members had drafted a proposal for allocation of raw materials with an eye to plans for an atomic pile in the United Kingdom. The 4 July 1945 agreement on allocation held for production during the war. Now that the shooting was over, the British proposed that all materials received since V-J Day and through 31 December 1946 be divided on a 50-50 basis between the United States, on one hand, and Canada and Great Britain, on the other. Criticism of American heavy employment of the scarce materials was implicit in the additional suggestion that immediate consideration be given to more economical usage.[29]

The American delegation made a counterproposal that all materials received and utilized by the several governments through 31 March 1946 should be considered as allocated to those governments. The United States had received the great majority of these supplies and of course did not wish to have to return a portion to its ally. After March, the Americans suggested that the United States should receive 250 tons of U_3O_8 per month to meet established plant needs. Any remainder might be allocated by the CPC in accordance with needs of respective programs upon written application. Groves was not about to let the British automatically receive even those ores not immediately required by American plants. Finally, all uranium recovered by the American forces in Europe during the war were to be allocated to the United States under the American proposal. Of particular interest were approximately 525-600 tons of ore which had been recovered in Germany and elsewhere, part of the ore which had been at Oolen when Germany invaded Belgium in 1940. When it was discovered in the closing weeks of the European war, it had been shipped to Britain for safe keeping.[30]

When these conflicting proposals were considered at the April CPC meeting, Halifax termed the British position as "an equitable and simple solution."[31] He saw the U.S. proposal as "inequitable" and unacceptable to his government. Britain was receiving little information and no material; yet she was financing up to 50 percent of the supplies received.

Groves contended that the principle of the Quebec Agreement and of present allocations was need. Unlike the United States, the United Kingdom had no current need for material to operate a plant; on the other hand, any diversion of the existing flow would force a partial shutdown of American operations. Halifax conceded to the *status quo* but only if a prompt decision were reached by a special committee appointed to study the matter. Clearly upset, he closed the meeting by stating that it left him with a "very uneasy feeling." While trying to work out full international cooperation, he feared the powers might impair their own.

The deadlocked CPC members referred the matter to their heads of government. The very next day Attlee telegraphed Truman that he was "gravely disturbed."[32] The course of discussion left the United Kingdom in a position inconsistent with the November Agreement that had been signed. He saw that document as meaning no less than "full interchange of information and a fair division of the material." The conditions of war had meant that technological and engineering information accumulated in American hands. "Full and effective cooperation" required that this now be shared. If the CPC could not arrange it, then the heads of states should issue instructions for such interchange.

Patterson did not sit still either. The secretary of war pointed out to Byrnes that while the short statement signed by the heads of state did call for full and effective collaboration, the longer Memorandum of Intention drafted by the same group at the same time specifically distinguished between basic scientific research and technical and engineering information. While the former might freely be exchanged, the latter was subject to *ad hoc* decisions by the CPC.

This was the position that Byrnes defended to Halifax on 18 April and Truman to Attlee two days later. Truman admitted not having prior knowledge of the November 1945 Memorandum of Intention or of the restrictive wording Groves inserted. But he assured his counterpart that he never would have signed the shorter declaration had he thought it could be construed as implying that the United States was obligated to share technical information. He thought it unwise for both countries for the Americans to assist the British in building another plant; indeed, he suspected current American public opinion would be against the building of another plant anywhere until the issue of international control were settled. The president's mind was made up, and he would not be lightly dissuaded. Though Attlee pressured Mackenzie King to support his position, the Canadian, presumably aware of the wording of both November accords, declined to do so. Instead, he merely reminded the Americans that Canada had given Americans full

access to developments at the Canadian Chalk River plant. All the negotiating left the Canadians irritated and disappointed; Pearson found
Byrnes ill-informed, and C. D. Howe noted that the United Kingdom
seemed to want control of both the operation and output from Eldorado; Canada, he thought, could not agree to this and in the absence of
an alternative might do well to withdraw from the Trust.[33]

The difficulties experienced with the British over shared information
naturally meant that the American response would be negative when
Prime Minister Spaak of Belgium requested permission to make some
statement about the disposition of Congo ores and the exchange of information promised by Article 9(a) of the 1944 Tripartite Agreement.
But if the Belgians were willing to be stalled, the British were determined to balance their own defeat on the information issue with some
sort of gain in the allocation of materials.

Groves was ready for them. When the British vigorously argued in
the subcommittee that all materials received since V-J Day be split 50-
50, the general categorically refused. His reasons were multiple. The
British were, of course, eager to build their supplies against the time,
perhaps three years hence, when they would need them for their proposed plutonium and uranium separation plants. They estimated their
need at 5,400 tons of U_3O_8. Groves saw U.S. operation and research
needs as being 3,060 tons for 1946 and at least 230 tons per month
thereafter. Yet the estimate of remaining high-grade reserves in the
Congo was only 7,700 tons. The 50-50 proposal, Groves wrote Acheson,
would decrease the present scale of operation of American plants. It
disregarded the principle of need and allowed the British to build
stockpiles for which they had no immediate requirement. Such even
division ignored the small amount of British contribution to the program so far.[34]

Any imminent war, Groves went on, implied an Anglo-American alliance, with the United States again carrying the major war burden (his
use of the word "again" here would no doubt have especially infuriated
the British had they known). If the two countries would be partners in
any military action, how could a duplication of plants be justified, given
the scarcity of materials? Groves was building a neat, if oversimplified,
dilemma. Either the British would be allies, therefore not needing
plants, or be neutral, or be enemies, in which last cases they should not
have plants.

The real purpose of the British claim, Groves thought, was "to build up a stock of materials to take advantage of potential commercial uses. . . . A decision to divide the raw materials on a 50-50 basis would, in effect, be penalizing the United States for using its raw material to insure peace, while permitting the British to stockpile their raw material for future peaceful uses."[35] Though Groves was no eager advocate for international control, he saw the opportunity for an additional argument that might influence his superiors. Would it not be unwise for the United States to weaken its hand by giving information or material to any nation while bargaining for an international control solution was still in progress?

The adamant position of the commanding general of the Manhattan Engineering District forced the British to a fall-back position. This granted the Americans all materials acquired prior to April 1946 (approximately 1,134 short tons of contained oxides), plus half of those delivered for the remainder of the year. The Americans would have to return nothing, could receive the materials captured by their troops in Europe, and be allowed to purchase the current Canadian production of 30 short tons of oxide per month. This would give the British 1,350 short tons for the year, along with 50 tons of Mallinckrodt oxide and 15 tons of uranium metal (the equivalent of 100 tons of contained oxide) which the Americans would ship to them. Because such an arrangement still left the United States 408 tons short of Groves' estimated need of 3,060 tons for 1946, Groves saw even this as "not fair to the United States because it is contrary to the principle of 'need.'"[36]

In retrospect it appears that Groves overreached himself in rejecting the British compromise. Such inflexibility was inappropriate at a time when the Americans, as much at Groves' insistence as at Bush's persuasion, had taken a narrow and firm position on information sharing. What, too, of British innuendoes that the Americans were spendthrift in their use of uranium? Halifax was incensed at the shameless extent of Groves' appetite for uranium and even more so by aspersions cast on Britain's ability to defend herself and maintain security. The ambassador argued fiercely with Acheson and at one point threatened to take responsibility himself for saying his government would no longer agree to shipment of the precious materials to the United States and would call for possible liquidation of the CDT if the Americans did not acquiesce in five days.[37]

Though Halifax did not realize it, and although petulant ultimatums can sometimes produce an effect contrary to that desired, Acheson and

Patterson both thought the British had a just claim and at the end of May were inclined to overrule Groves.[38] They apparently did, for the British plan was accepted by the American subcommittee consisting of Acheson, Bush, and Groves. Groves did get agreement that no more than 1,350 tons of oxide be allocated to either country for the remainder of the year. He also achieved agreement that this arrangement would not prejudice distribution in coming years; in fact, of course it did.

The Americans acquiesced because they did not believe the British would be satisfied with anything less. Failure to agree might endanger the future of the Trust and the so far successful collaboration on raw materials. No doubt the thought of South African ores lay somewhere in their minds. Groves did win a point in persuading his colleagues to question the advisability of building atomic plants in Britain, where they might be neutralized, destroyed, or captured in the first days of hostilities.[39]

There remained the matter of who should pay for what. The British were prepared to pay half the $885,468 cost of the approximately 293 short tons of U_3O_8 purchased prior to V-J Day, for they had profited by the weapon which accelerated the end of the war. Between V-J Day and 31 March 1946, 850 short tons of contained oxide had been received, costing $2,582,260. Deliveries from the Congo had increased dramatically since then as new production efforts and machinery took effect. From 1 April until the middle of July 1946, 1,969 short tons had been delivered at a price of $7,113,956. The British wished to pay for only the ores they had received since 1 April; this was agreed upon at the end of July.[40]

The arrangement brought with it a windfall for the British of which both governments were aware. Britain continued to pay for half the cost of the materials acquired by the CDT in sterling and then was reimbursed in dollars by the Americans for the extra 40 percent of uranium that went to the United States above its regular share of 50 percent allotted by the new agreement. At a time when dollars were at a premium in the world market, the arrangement assured the British access to a goodly number of dollars. Union Minière and other producers would of course have liked to have received those dollars but had to accept half their payment in sterling. The arrangement caused the United States no harm and helped make the allocation arrangement more palatable in London.[41]

The McMahon Bill and the Resignation of Groves

Collaboration on the acquisition of raw materials thus continued, but information exchange remained at a halt. The deadlock ostensibly resulted from the conflicting British and American views encapsulated in the two November Agreements. Though the short one signed by the heads of states called for "full and effective cooperation," the longer specifically distinguished between basic research and applied technical knowledge. This distinction would have been of no great significance had the CPC been likely to grant *ad hoc* approval of the exchange of pertinent bits of information. But it would not. American opposition was too strong, and the three American votes could produce a stalemate even if the Canadian representative sided with the two British delegates.

Concern for the international control talks affected the American attitude. Less important was the feeling occasionally evident in Groves' outbursts to the effect that the United States had made by far the greater effort and had the right to keep its own discoveries. Some American military wanted to keep vulnerable plants out of England. Above all, the chief influencing factor was security. After the exposure of the Gouzenko spy ring in Canada, there simply was no likelihood that the United States would be amenable to further information sharing. Groves and others had been aware of various Soviet spying efforts; this was one reason that Groves arranged in December 1945 for the creation of a special Combined Intelligence Section of the CDT which dealt with all aspects of atomic intelligence while working through existing agencies in the three countries.[42] But the Alan Nunn May affair was one of the first clear indications of successful Soviet penetration, and the weak link had been British.

Meanwhile the McMahon bill was making its slow way through congress. By 1 April the section on dissemination of information had been retitled "control of information," the distinction between basic scientific and related technical information deleted, and the entire emphasis shifted from releasing to restricting information flow. Though much politicking remained to be done, by late spring the act was fairly well framed; passed by congressional action, it was signed into law by President Truman on 1 August 1946.

The bill's detrimental impact on Anglo-American relations appears to have been only slowly acknowledged in both the state department and the foreign office. The clash between the military and congress was reason enough for the state department to keep its distance from what was primarily a piece of domestic legislation. The reluctance with which the department later faced the disagreeable task of informing key senators of the Quebec Agreement and the volcanic eruption that followed attest to this.[43] Moreover, the early versions of proposed atomic energy bills seemed to pose no problems, as the British embassy reported. It was after the discovery of the spy ring that important changes were made. These came quickly and any attempt to oppose them would have appeared as betraying U.S. security. One authority has described the scene in London when the senate quickly approved the bill in June as "sudden panic."[44] Anderson rushed to Attlee; the notion of an appeal to Truman seemed plausible. The embassy in Washington, however, cautioned that there was little chance of executive intervention now that some legislative agreement had been reached after months of infighting. A telegram would be inopportune.

This advice did not dissuade the prime minister from wiring, but it probably caused his language to be indirect. Attlee argued that had the United States not agreed to take on the building of the bomb, Britain would somehow have done so; she should not be penalized. He avoided an accusatory tone or assertion of bad faith, yet there was a plaintive and even reproachful flavor to his lengthy message. Why should collaboration continue on materials and not on information?[45]

Truman did not reply. In December Attlee jogged him again. By then passage of the McMahon Act and the creation of the U.S. Atomic Energy Commission made the situation all the more complicated. While the bill restricted information sharing, it did require that congress be informed of all obligations regarding materials and the like. Yet, up to that time, the American congress and the British parliament were still unaware of the Quebec Agreement, the Hyde Park *aide-mémoire,* or the November Agreements. The proposal for international control of atomic energy was severely bogged down in the United Nations, and events in Turkey, Greece, Germany, and Czechoslovakia forcefully indicated that the dream of postwar great power cooperation was over. With all these concerns, it is not surprising that Truman put the issue of exchange on the back burner.

The British, however, were having technical difficulties in planning the construction of the plants which the Americans so strongly, but for the most part tacitly, opposed. American technical assistance could save six months and much cash. Though the lack of cooperation and passage of the McMahon bill had caused the British to conclude "that at heart America simply wanted to retain her existing atomic monopoly," they thought effort at breaking the log jam worthwhile.[46] Attlee did not blame Truman. He saw that the president had to pay a price to keep basic control of atomic energy in civilian rather than military hands; the blame, he thought, lay with congress.

An attempt to work through Eisenhower, who the British considered as friendly and not opposed to atomic plants in Britain, was canceled in March 1947 when Ike tipped the British off that the chiefs of staff of whom he was chairman had informed the state department of their opposition to such construction in the United Kingdom. The general attributed this posture not to distrust in British precautions but to concern for future conditions. The British, however, saw this as meaning the Americans, concerned by the growing economic crisis, strikes, and disorder in Britain and her inability to maintain commitments in Greece and Turkey, feared that their country might experience instability and a further shift to the political left.[47]

A different approach was necessary, and in time it was to bear fruit. On the first of February, Roger Makins, the British plenipotentiary who served under Groves as deputy chairman of the CDT, called on Acheson at his home. The failure to reach agreement on atomic policy by the UN, the new U.S. legislation, and the succession of George C. Marshall to the post of secretary of state on 21 January 1947 suggested possible movement on the issue of information sharing. Makins pointed out the irritation of technological delays and the psychological difficulties caused in London by the belief that the United States was willing to cooperate in the field of raw materials, where the United States might gain substantial benefit, but not in other fields where the benefit would flow to Britain.

Though Makins stated that his suggestions were purely personal, it was obvious they had been carefully prepared. Could the United States provide information on just twelve to fourteen specific points where help was currently needed? Could information be passed on which had been obtained prior to passage of the McMahon Act? What of merging military applications of atomic energy with ongoing exchanges of other defense information? Could the tone of the American reply to Attlee's note be conciliatory, and could the United States give a blessing to the

U.K. atomic energy program, to dispel the impression of hostility toward it? Could there be better exchange of unclassified knowledge?[48]

Acheson withheld formal reply, but his attitude was not positive. Yet he could not afford to be entirely negative, for he also now needed an agreement. Word of the Quebec Agreement would soon have to be given to congressional leaders, as the McMahon Act required. Failure of the November 1945 attempt to rewrite that accord meant that the clause requiring British consent prior to any use of the bomb was still in effect. No crystal ball was necessary to guess what congressional reaction to that news would be. Acheson therefore inquired if existing arrangements—not meaning the Combined Policy Committee or Combined Development Trust—could be terminated. To this Makins assented, as long as termination was part of a larger settlement.

Implementation of the McMahon Act, which went into effect 31 December 1946, could not await satisfactory arrangements with the British. The creation of the civilian Atomic Energy Commission, which took over the American side of the CDT, left Groves with reduced influence. No longer able to serve as secretary to the CPC or as chairman of the Trust, he was made a member of the Military Liaison Committee. At the end of June 1947 Eisenhower explained the appointment to Lilienthal. Groves had been included in order not to alienate his supporters and in order to "pump him dry" of any helpful information he might have. The chief of staff admitted that Groves had enemies in the Pentagon and that he lacked an understanding of how to get things done without "humiliating people and making enemies." Eisenhower assured the AEC chairman that he would take Groves off the MLC if he caused trouble. The general went on to state that the British blamed Groves for the denial of information exchange. In his diary Lilienthal quoted Eisenhower as saying:

> They [the British] realize that the McMahon Act forbids you to exchange information or supply new information, but they blame Groves for going around behind their backs and having that provision against exchange of information put into the McMahon Act, even pointing to the line which singles out "industrial uses," which shows on its face that whoever got that line inserted had the background which only Groves and two or three others had. They feel he has overtraded and are deeply upset by it.[49]

By the end of July, Lilienthal, who thought many of Groves' comments shallow, was fed up. "After sniping at us, sneering at us and running us down, he may find this meeting will ... [require a defensive] role. We have taken all the kicking around we intend to take."[50] In September Lilienthal was pushing for Groves' removal from the Armed Forces Special Weapons Command and, if possible, from the MLC. Eisenhower was amenable, but his views were at first overridden by the other chiefs of staff. Conflict continued to build among Lilienthal, the commissioners, and Groves, especially over the issue of military custody of atomic bombs. At the beginning of 1948 Secretary of Defense Kenneth Royall was discussing with Eisenhower ways of easing Groves out when Groves—aware of the mounting pressure—conveniently decided to retire on 29 February in order to enter private business.[51]

The new Atomic Energy Commission members immediately questioned the continued operation of the Combined Policy Committee under the new U.S. law and pressed the state department to dissolve the CPC or at least inform congress of its existence. This was partially done in May, when the Joint Committee for Atomic Energy Affairs, drawn from both senate and house of representatives, was told that U.S. efforts depended on Belgian ore and that the British were currently receiving half of that ore. Concerned by the recent Communist withdrawal from the Belgian government, Acheson a week later urged secrecy on the group because the Communists had been "vigourously pushing for abrogation of the contracts," and any publicity would strengthen their hand, endangering the U.S. atomic program.[52] He then told the Joint Committee of the Quebec accord, the CPC, the Trust, and, in general terms, of the several agreements to procure uranium and thorium.

All this did not sit well with members of the Joint Committee and especially with its conservative chairman, Senator Bourke B. Hickenlooper of Iowa, or with Senator Arthur H. Vandenberg of Michigan, who was also chairman of the Senate Foreign Relations Committee. Vandenberg took up the Quebec matter with Truman. Hickenlooper described himself as "shocked and astounded," and called the Quebec Accord "ill-advised."[53] The Iowan was further upset that Britain was stockpiling its share of Belgian ore and that other sources within the Empire were not available to the United States. He saw such stockpiling as unnecessary and a target attracting invasion. Hickenlooper insisted that this uranium be brought to the United States, perhaps as collateral for the large loans the British were negotiating, but primarily for security. The Quebec Agreement should be rescinded. If the United

States was expected to pull British chestnuts out of the fire (as in Turkey and Greece, as well as her domestic economy), then the country should have all the implements it needed. He would oppose any further aid to Britain until the two matters were solved.

The notion of curtailing aid unless uranium were handed over was aimed primarily at Britain. But it had implications for Belgium and for the entire European Recovery Program. In a September meeting with the secretaries of the armed forces, George Kennan was able to carry the view of the state department that such a linkage might backfire and should not be used.[54] Yet the need for an agreement was all the more pressing. Careful review of oxide supplies made clear that, without access to the British stockpile and to the ores from the Congo going to that stockpile each month, U.S. plants would soon operate only well below capacity. Already the American joint chiefs of staff had expressed their concern, perhaps irritation, that unprocessed ore was accumulating in England. Though they saw the danger of a British plant diverting material from U.S. production and being too near an enemy, their overriding concern was that the ore be quickly processed, so that fissionable material would be available in an emergency.[55]

The Modus Vivendi *of January 1948*

With Acheson returned to private law practice in mid-summer 1947, Marshall set Kennan of the Policy Planning Committee to work along with Gullion on a program for negotiations. Their suggestions were presented on 24 October. The CPC and CDT should be maintained with any necessary minor alterations, but all other related wartime agreements should be suspended. The three governments—the United States, the United Kingdom, and Canada—would endeavor to control uranium in their own territories, the Commonwealth, and elsewhere as determined by the CPC. All supplies would be placed under the CDT, which would allocate them according to need. What was not used in current operations would be stored in the United States, and Britain would ship what she presently did not need across the Atlantic and would do no further stockpiling. There would be full and effective cooperation in basic research, while exchange on technical aspects would be regulated by the CPC on an *ad hoc* basis. This was essentially Groves' old Memorandum of Intention formula. Information would not be shared with other powers without prior consent. What of the McMahon

Act? The government would go to congress to obtain wider authority to exchange information when it would serve national security.[56]

These recommendations were considered by the American members of the CPC. Marshall feared that the strain among the Americans, British, and Canadians might permit pressure to build on Belgium to sell elsewhere. Bush pointed out that the progress of Canadian and British projects was such that they now had information to give to the Americans. Lilienthal and Herbert S. Marks, general counsel to the AEC, conscious of their legal mandate under the McMahon Act, warned against linking the issues of information and materials in any talks with the British. But Kennan commented that it was "not feasible" to expect to get materials while giving nothing in exchange. The failure of negotiations in the UN meant there was no longer reason for waiting. Lilienthal feared delay in getting any new legislation passed. He and Marks thought that

> if it was established that their [existing relations with Britain and Canada] continuance was in the interest of our own national defense and required . . . exchange of information, . . . a sound argument could be made that such arrangements were permitted by present law.[57]

Lilienthal wrote in his diary of his annoyance that "the State Department staff people, Kennan and Gullion, keep pressing for a broad approach—something that smacks of an alliance. I shall do everything I can to discourage and prevent this. . . ."[58]

A flurry of additional meetings followed. Joseph Volpe, now on the U.S. Atomic Energy Commission legal staff, took the task of deriving a formula which would allow linkage of materials and information exchange by satisfying the AEC representatives that such exchange was legal because tied to national security. Debate on the issue faded, as it was learned from informal contacts with British scientists that their greatest needs dealt with health and radiation safety and with the most advantageous way to use the Canadian Chalk River plant. They were also busy in the field of medical isotopes. Exchange in these areas did not seem threatening and would benefit both powers.[59]

Moreover, Bush saw the need for the assistance of the English intelligence net regarding Soviet progress. Kennan's view was that just reopening negotiations would improve the situation, as the British would be reassured that the special relationship of old on atom matters still existed, despite the fact that exchange issues remained unsettled. The issue of South Africa and the strength of British and Canadian irritation with U.S. stalling and the existing one-way flow of information from Britain to America were brought forward. Nor could the Americans overlook British influence in Belgium.[60]

The strength of the British bargaining position in terms of materials impressed Vandenberg and Hickenlooper. Both were annoyed that British stocks were lying unused while American production was slowed. Vandenberg made clear that, if the British proved obstinate, he would try to see to it that any "further loan assistance to the British took account of their failure to meet us half way." He considered it "inadmissible" that Britain "should hoard uranium, making no use of it at all, when it might be made into weapons for protection of the democratic world."[61]

It is noteworthy that the two senators did not balk at the presumed possibility of renewed information exchange which relied on careful definition of national security and some stretching of the wording of the McMahon Act. Their chief concern was getting uranium and eliminating the British veto on use of the bomb. Moreover, the state department's negotiating position, as carefully explained to them, was initially only to explore the areas in which the British desired information so that some decision could be reached as to the effect that cooperation in these areas would have on American security.

Under Secretary of State Lovett remained more than nervous about the possible linkage of atomic energy problems with the debate on the European Recovery Program. This would corroborate Communist propaganda that the Marshall Plan was a means to a U.S. atom monopoly, lead to unfortunate public debate of U.S.-U.K.-Canadian agreements during the war, and possibly endanger passage of the European Recovery Program. He asked his ambassador in London to explain to the British the need to hasten talks to meet the effective deadline of 17 December, which Vandenberg appeared to have set, after which he might speak up in the senate.[62]

The British, prodded by their own concern for information, their desire for a reestablishment of the old relationship, perhaps also by the growing tensions in Europe, and encouraged by what they considered the new assembly of pro-British personnel in Washington, responded well to informal messages about the trend of American thinking. The three key areas of concern—information exchange, materials, and bomb control (the Quebec Agreement)—were formally raised at a major meeting of the Combined Policy Committee on 10 December.

Subcommittees were appointed to consider each of the first two issues. That dealing with information promptly identified nine areas for a beginning exchange list. These ranged from health and safety to isotope research, detection of distant explosions, extraction chemistry, the design of natural uranium reactors, and general research experiences with certain low-power reactors.[63]

The materials sub-group had more difficulty. It did agree on estimated needs and supplies available. These did not match, even when American estimates for low-level, rather than high-level, operation of their plants were used (the British submitted only one estimate level for their more limited operations). Only if the bulk of the 3,250 tons of uranium stockpiled in Britain were used could a more appropriate balance between American needs and supply be achieved; further, the British would have to forego a portion of their monthly shipments from the Congo. The British had already made clear that they would find it difficult to agree to any proposal which implied that the United Kingdom was not a safe place for stocks, but the issue of operating needs put the debate on a different footing. The supply and demand dilemma was referred to a new subcommittee.

To ease the susceptibilities of certain Belgians and in accord with a suggestion by Sengier that the highly capitalistic sounding word "Trust" be dropped, the CPC agreed to rename its purchasing arm; eventually the title "Combined Development Agency" (CDA) was chosen. It was also decided that any new Anglo-American-Canadian agreements were simply to be entered in the CPC minutes to avoid any implication that they were treaties requiring registration with the United Nations. Thus was the stalking horse which had earlier been used so prominently to avoid settlement put out to pasture.[64]

In the materials negotiating group the Americans asked for all of the 1948 and 1949 Congo production, plus appreciable portions of the British stockpile. Enough would be left to supply British activities through 1949. Such allocation, though an imbalance in tonnage, would supply the needs of both powers equally and leave them with equivalent reserves. But Makins and British Ambassador Sir Gordon Munro were authorized to deal only with the Congo production for 1948, which they were willing to grant in its entirety to the Americans.[65] Though the British negotiators seemed sensitive to the American concern, they would not budge on the stockpile issue. Kennan for his part was careful not to offend British Foreign Minister Ernest Bevin's tenderness about implications that materials in England were insecure.

The deadlock continued. Kennan contacted Chip Bohlen, an experienced department of state counsellor then in London and an old friend, on the evening of the 17th. It was plain to see, he said, "that difficulties stem not from the British delegation here but from the Cabinet in London, where decision appears to be made, largely on emotional grounds, that none of the supplies on hand in England are to be given up."[66] Kennan believed Makins was impressed by the good will and logic of the Americans; he would be returning to London to ask the cabinet—actually the Official Atomic Energy Committee—to reconsider. Kennan was sure Makins was aware of how hard it would be to defend the British position before congress. Kennan urged that the American representatives in London be sure the British government understood the reasonableness of the U.S. position, its justification in terms of need, and "that unless there is some give on their part we can be faced with a situation in Congress which could have appalling consequences."[67]

The message got through, and the British government relented. The old resentment of U.S. monopoly of uranium in the early years of the war had decayed, and there were memories of several years of fruitful collaboration. Though South African ores were not yet available in quantity, they soon would be. British ability to serve as custodian of the ores was not being questioned, nor apparently was the existence of atomic plants in Britain.The British recognized that Vandenberg and Hickenlooper could make the issue one that disrupted and delayed passage of the Marshall Plan, funds from which were vital to British recovery and for the rehabilitation of European economies on which British trade and thus future economic welfare depended. The London leaders refused to bridle at senatorial pettiness, instead opting for the higher road of statesmanship. The abolition of the restrictive Clause IV of the Quebec Agreement would be achieved, and perhaps more exchange with Commonwealth countries would be possible.

Only one voice, that of Vice-Chief of Air Staff Sir William Dickson, questioned the abandoning of the Clause II Quebec provision for mutual consent prior to use of the bomb. Makins also apparently pointed out that by the end of 1949 American uranium reserves would be down to a three-week supply, and, though the United Kingdom claimed the stockpile in Springfields, Lancashire, in fact it was held on joint account. The Canadians were blunt in their opinion that the British position was unfair and indefensible.[68] Then, too, the growing Soviet threat also fostered an inclination to assist American bomb manufacture, a feeling that had not been so strong eighteen months earlier.

Helping the Americans in the long run helped the Western alliance. Moreover, the British would be making informational gains. The chance for agreement and to work with preponderantly pro-British personnel in Washington too might be fleeting, Makins argued. It would be best to reach a solid agreement on the problem "before the rats can get at it."[69]

Conclusion

Despite the reluctance of the American military to share information and the near prohibition by congress of such action, the British had managed to crash through both obstacles. No doubt the reduction of Groves' influence eased their task; yet it was the bomb clause of Quebec and the scarcity of supplies that gave them leverage to move the leading senators and the American military. Vandenberg for his part was no tyro. He was working to get movement out of both the American bureaucracy and the British. His threat may have been empty, but no one could be sure. Though Kennan and Lovett's reasons for not linking the European Recovery Program with atomic issues were sound and the senator might have yielded to them in terms of public statements, yet he could and did cast the shadow of such a linkage in private discussion; blockage of the legislation in committee by whatever means had to be feared. In fact there was linkage of the information and materials issues, all the careful statements to the contrary. Interestingly enough, the conservative guardians of congressional prerogatives scarcely blinked at the circumlocution of the provisions of the McMahon Act. Flexibility had to be shown. It was Lilienthal of the AEC who proved most cautious on this point.

In retrospect, the Americans may have regretted acquiescence to the 50-50 proposal that allowed the British to establish their stockpile. Yet the logic of that decision, including future need for South African ores, still stood. Relations had been bad enough. What if there had been failure to compromise on this issue as well? It is safe to say that the continuing cooperation on supplies while information exchange had broken down was probably the single most important factor in paving the way for new agreement. The Americans estimated that over the next two years the British stockpile in excess of operating needs would be reduced to one-fifth its current size. Vandenberg and Hickenlooper were pleased, though the latter remained concerned that a leftward swing of the British government might bring danger of the supplies being diverted or sold.[70]

The January 1948 *modus vivendi* provided each of the powers what it wanted. Nine specific areas for information exchange were settled upon, and a new arrangement for uranium allocation accepted. Once the English and the Americans both desired a deal more than not, agreement had been reached despite problems actually greater, because of passage of the McMahon Act, than those which existed when exchange broke off. Yet the long shadow of the secret diplomacy conducted at Quebec had not been completely dispelled, and the thinking which produced it still had its influence. Agreement there had been reached more by the determination of the top political leaders for cooperation than by mutual willingness at the lower levels where the accord would be implemented. The Americans' desire to keep the keys to production of the ultimate weapon in their own hands and land continued, as did the British wish to have access to the new instrument which gave the seal of great-power status. The Americans did not accept the British contention that the price for such access had already been paid in full by an overall greater contribution to the World War II effort by the British than the Americans. The situation was not improved by the attitudes of the scientists, who, on the one hand, favored wide exchange of basic ideas within the scientific community and, on the other, jealously protected the prestige conferred by their personal discoveries.

In the discussions leading to the Quebec Agreement a major American concern had been that the British would take commercial advantage of knowledge gained from research in the United States, while little opposition was voiced to the exchange of knowledge for military purposes. By January 1948 and the negotiations resulting in the *modus vivendi*, the situation was reversed. Why? The answer may lie in the change of American perception of the British Empire and in the deep-rooted distrust and misunderstanding in the American establishment—whether military or political, east coast or mid-west—of left-leaning governments, a distrust greatly augmented by the spread of Soviet domination through Eastern Europe.

In 1943 Britain was still viewed as *the* great empire, with wide power and economic connections. In the middle of 1947, as the United States settled back into "normalcy," it looked across the water to see a Britain on her knees, wracked by economic woes, ruled by a labor government unable to repress rampant strikes, defending a falling currency unsuccessfully, dependent on American loans, losing grip on her empire, and defaulting on military commitments in the Mediterranean. British commercial competition no longer seemed a threat, and indeed no effort was made to enforce Clause IV of the Quebec Agreement. Britain's weakness, together with her leftist government and the exposure of the Gouzenko spy ring, made her a military liability in the eyes of some Americans. If their reaction to the spy discovery verged too close to the hysterical, the failure of the British to comprehend the depth of the American reaction reflected dullness.

The documents suggest that, despite those who argued to the contrary, the Americans' interest in achieving international understanding on atomic matters with the Russians was genuine and the fear that further development of the special atomic relationship with Britain would jeopardize such understanding was legitimate. Tremendous efforts at secrecy were made which delayed information exchange. Despite the secrecy, presumably many of the key exchanges were known to the Russians through the spy Donald Maclean, then serving as first secretary of the British embassy in Washington and as one of the British atomic energy specialists. It will be a long time, if ever, before the West knows how Stalinist Russia interpreted Maclean's reports. But it may be guessed that enough interchange occurred to stir Soviet suspicions, while the effort to avoid arousing the Russians prevented that information exchange from being meaningful to the British production effort.

From the British standpoint, much of what transpired in the postwar period was not comprehensible. Promises seemed to be made and not kept. To the official historian of the United Kingdom Atomic Energy Authority, it seemed

> extraordinary that the McMahon Bill, with its complete disregard for solemn commitments to the British and Canadians, both in writing and in spirit, should have been allowed to reach the statute book without, as it seems, a murmur from the Administration.[71]

To Americans, acquainted with the nature of their politics and the mood of congress at that time, it was all too comprehensible. As Acheson commented, it was the price paid for the American style of government and the loose way some things were done in the brash and still young country.

Margaret Gowing has suggested that:

> In the long history of strange atomic energy agreements, the *modus vivendi* emerges as the strangest of them all. . . . The British . . . gave the United States negotiators credit for understanding what they were doing, and again it was to transpire that they did not. These unpleasant revelations lay in the future. In the New Year of 1948 no one seriously doubted that the egg was a good one. Only later did the British find that it was addled after all.[72]

The problem, as Gowing sees it, was that the individuals favorable to Britain, such as Kennan, AEC manager Carroll Wilson, and Lilienthal (in this case she may have overestimated the Anglophilia), were not sufficiently representative of the American bureaucracy. Makins recognized that there would be mice—or rats—who would be quick to nibble at any accord. He saw the *modus vivendi* as a gamble worth taking in the hope that it could be strengthened, rather than weakened, as time passed. His recognition that delay might abort its birth indicates that he knew it did not have strong support.

The *modus vivendi* made sense, but more from an American perspective than a British. The Americans wanted to get something done about the Quebec Agreement and in time to avoid embarrassment of the European Recovery Program debate. One could start from a small area of information exchange and build from there bit by bit. The British argument, however, was based not on pragmatism but principle; a bargain had been made and should be kept, regardless of changes of circumstances such as domestic legislation or discovery of spy rings. Similar differences of perception had arisen in the negotiation of the 50-50 accord. The British argument was based on a principle of fairness or equity, while the American case was based on a pragmatic argument of need. These differences of approach were in turn based on differing assessments of the role each power would play in the postwar world. The British clung to their concept of a great power acting as the embodiment of principle, while the Americans saw the British stand a luxury made possible only by the Americans' pragmatic attention to the security of the West. It was the two countries' recognition of their mutual need for each other and, as Gowing points out, the British determination not to let their American cousins get away with unjust behavior, that kept the diplomats talking when others might have abandoned the effort.

The issue which ultimately brought the sharpest criticism in Britain was only briefly considered at the time. This was the abandonment of the right of consent to any use of the bomb and even of any mention of prior consultation as had been provided in the aborted February 1946 attempt to rewrite the Quebec Agreement. The matter probably did not receive much discussion because it was patently clear on both sides of the Atlantic that Clause II had to go if anything, including the Marshall Plan, were to move forward. The British made a virtue out of the necessity of abandoning that clause and gained the dropping of Clause IV. Had they not agreed, there is little doubt that Vandenberg, Hickenlooper, and the American congress would have seen to it that the Quebec Agreement was abrogated in any case.

The long loss of information exchange prior to the *modus vivendi* was regrettable in terms of the building of the British program, but less so in light of the leaks that did occur. Both British and Americans knew that the British could succeed; the irritation lay in being forced to reinvent the wheel at considerable cost. In one instance, the production and separation of plutonium, the British benefitted; because of lack of access to American plans, they came up with a process which was more efficient and in the long run helped to erase the uranium shortage. Nor did the relinquishing of sole claim to the uranium stockpiled in England ever harm the British, for in fact ore never had to be shipped from those stocks to America.[73] On the whole, however, the loss to the British in time, monies, and scientific effort, not to mention prestige, was great.

Confusion and infighting in Washington severely delayed the negotiations. The conflict between the war department and the department of state, despite Eisenhower's efforts to mediate, was appalling. So too was the less open but always more than latent conflict between congress and the president and between members of congress and the military or diplomatic corps.

Some of these conflicts were to be expected as congress attempted to wrest influence over foreign policy from an executive and military that had grown increasingly imperial through the crises of depression and world war. Ironically, the squabbles that ensued did more to justify than to decry such sweeping presidential acts as those of Roosevelt at Quebec and Hyde Park. They also showed the dangers of leaders taking steps when they do not have the time or energy to persuade their lieutenants of the wisdom of following in their tracks. The events associated with the uranium hunt both before and after the *modus vivendi* was reached illustrate these problems vividly and reveal the lack of unity so desperately needed in the Western world as the Soviet challenge mounted.

6 · *The Difficulties of Sharing*

Though the *modus vivendi* was the product of painstaking diplomatic work, it was more a papering over of divergences than a monument to a meeting of minds. In the long run its ability to function would be impaired as much by differences between the U.S. congress, state department, and military as by those between the United Kingdom and the United States. These first variances stemmed from multiple factors, including the "red scare" hysteria. Not least among them was the heritage of the manner in which the Manhattan Project, the uranium hunt, and promises of information sharing had been kept primarily within the military. Nor was the continuing influence of Groves' reluctance to share information with civilians negligible. As in the creation of the *modus vivendi*, it would be American desire for more uranium that would bring adjustments in governing regulations, adjustments which would expand avenues both for information sharing and proliferation of nuclear capabilities by diplomatic bargaining.

Interchange in Trouble

The technical cooperation envisioned under the *modus vivendi* initially proceeded rapidly and well. Then in the spring of 1948 it suddenly appeared that certain Americans were not at all sure they did in fact want exchange, or at least as much as the limited agreement allowed. The rats were beginning to nibble. Exchange was delayed as the AEC ruled that each request for information had to be judged individually as serving the common defense and security. This already marked a separation from the British understanding that once the argument of common defense had justified the general agreement, then individual items required no more debate. Above all, the production facilities at Hanford, Washington, which the British most desired to see, remained off-limits. What information was received seemed patchy and the experimental data thin; the British began to wonder if the Americans really were as far ahead as they purported to be.[1]

A report from an American scientist in England triggered the first major reaction. Visiting in late May 1948, Walter H. Zinn found well-constructed reactors and considerable progress in the production of plutonium. Financier and former Rear Admiral Lewis L. Strauss of the AEC was distressed. He had been reluctant to agree to limited information exchange in the first place, and now it appeared that the British were on the verge of producing weapons. What information had the British given the Americans to match the obviously valuable help they had received? Strauss and other commission members had known since 19 March that the English were working on atomic weapons, but Strauss seemed to think that critical knowledge of plutonium had belonged solely to the United States. Although assured that this was not the case, that the British had enough knowledge of their own to make the bomb, and that plutonium was produced by any reactor, Strauss remained disturbed.[2]

Strauss' concern regarding plutonium was stimulated by his knowledge that the element was a key fuel for the American bomb production program. Indeed, the first atomic bomb, detonated in the Trinity test in the American desert in 1945, was made with plutonium. Essentially a man-made element that scarcely appears in nature, fissionable plutonium is produced by bombarding U-238 atoms with neutrons from a small amount of fissionable U-235. While the AEC plant at Oak Ridge produced U-235 by chemical processes, huge reactors at Hanford, Washington, "cooked" plutonium.

In August 1948, just as the British were asking for expansion of the areas of information exchange, Hickenlooper questioned whether the metallurgy of plutonium fell within the bounds of the *modus vivendi*. After contacting Strauss, he became concerned that exchange was wider than he realized. His agitation increased when he learned that further interchange on plutonium was about to occur. He and Vandenberg called on Secretary of Defense Forrestal to protest. Hickenlooper said he thought the British were interested only in industrial power; now their work on plutonium showed interest in bombs. Though Bush and others repeatedly pointed out that this was not news, that the British had made their intentions clear and all concerned had been informed in December, the chairman of the congressional Joint Committee on Atomic Energy would not be calmed. On his insistence the plutonium discussion was aborted.[3]

On 1 September, ignoring informal signals that exchange, especially related to weapons, was in trouble, the British presented a list of topics for interchange. A more formal and extensive warning given to the head of the British scientific mission by Carroll Wilson and Donald Carpenter, an assistant to Forrestal for atomic energy, did not reach London in time to head off the new approach, which had been in preparation for some months. The *modus vivendi* discussion had not touched on the issue of atomic bombs, apparently at American advice. Yet, as the British requests for uranium and research announcements made clear, this was the first goal of the British program. In London the anomaly seemed ridiculous, and Air Marshal Lord Tedder was disturbed by the failure to invite British representatives to bomb trials at Eniwetok in April 1948 as compared with those at Bikini in 1946. He saw the need quickly to learn more about the weapon that was supposed to become the key to his nation's defense. The British chiefs of staff in conjunction with Minister of Defense A. V. Alexander had therefore prepared the proposal for Forrestal's consideration.[4]

Alexander's covering memorandum argued that good defense required both countries to be strong; that in turn required close cooperation. Both powers would soon have the bomb; as it was central to their planning, they should concert their thinking on its development so that design of equipment and tactics would correspond. The British also expected to make important contributions as their program developed.[5] Though they might have withdrawn their proposal had Forrestal hinted it was ill-timed, its very presentation was indication of the urgency of their need. Whereas formerly the British saw the state department as more sympathetic to exchange than the military, now that congress was watching the diplomats so closely the American military were seen as more favorable.

Forrestal was frank, pointing out that there was concern over the vulnerability of any bomb production plant in England and whether the great sums should be spent, given the amount of dollars America was providing to aid Britain under the Marshall Plan. Admiral Sir Henry Moore, the head of the British military mission in Washington who carried the British message, assured the secretary of defense that these points had already been considered and that his government was determined to move ahead. Forrestal told Moore not to expect any prompt reply to his inquiry. This was in part because the coming elections might bring a change of personnel and viewpoints in Washington, but more importantly because there was no consensus in the U.S. leadership as to whether it should enter into full partnership with the British on atomic weapons.[6]

The reaction to the plutonium affair was to reduce exchange. The Military Liaison Committee established as part of the Atomic Energy Act to work with the Atomic Energy Commission became more of a watchdog. The MLC was supposed to approve the appropriateness of any specific exchange in each of the nine areas identified in the *modus vivendi*. In the weeks following the row over plutonium, the MLC became increasingly restrictive, while most of the members wondered whether exchange was becoming "so niggardly and reluctant that neither the spirit nor intent of the *modus vivendi*" were being carried out.[7] What, too, of the impact on allocation negotiations for 1950 and 1951, Arneson warned Lovett. A falling out on exchange "could very well sour" negotiations with the South Africans. His position paper concluded:

> In view of the high raw materials stakes involved—let alone the question of honoring the spirit and letter of our undertakings—it is recommended that the Department of State should throw its weight in favor of a more relaxed method of cooperation within the nine agreed areas, without prejudice to the larger issue of whether the areas should be extended to include the information requested by Admiral Sir Henry Moore on September 1.[8]

Divergent opinions on atomic policy continued to course through Washington. The elections of November 1948 brought little clarification. Truman remained in office, while the shift of political balance in the new congress between democrats and republicans introduced new chairpersons for all committees. For the Joint Committee on Atomic Energy Affairs, the new chairman would be Senator Brien McMahon.

A New Proposal

Some mechanism for resolution of differences among U.S. officials had to be found. In January William Webster of the Military Liaison Committee arranged for an informal gathering at Princeton, away from the currents of the capital. Included were representatives from the AEC, the department of state, and the military establishment, plus scientists James B. Conant and J. Robert Oppenheimer.

The supply situation by now looked less gloomy. Congo production in 1948 had been sufficient to meet U.S. needs without reduction of the

British stockpile. It was expected, however, that needed developmental work such as the opening of new shafts at Shinkolobwe in 1949 would mean that the British would have to ship between 600 and 1,000 tons of contained oxide from their stores. This depletion was far less than originally anticipated under the *modus vivendi* or demanded by Hickenlooper. It was hoped that the newly developed Redox method of efficiently reprocessing uranium wastes could be brought into large-scale utilization by 1952; if this could be achieved, then supplies from the Congo, Canada, the United States, and South Africa would meet everyone's needs until 1955. It was recognized that the British desired their own weapon for reasons of freedom of action, prestige, and uncertainty about U.S. policy toward the United Kingdom on atomic matters. American objections to plants in the British Isles remained, due to concern for loss of their production during a conflict because of their vulnerability, their less economic use of uranium, and the general diversion of alliance resources. British ability to produce a bomb was not questioned; the issue was just one of time and cost.

Taking these and other points into consideration, the group meeting at Princeton recommended full exchange of information, even on atomic weapons, creating what would basically be an integrated effort. Close consultation should be the rule, taking into account several principles. The most salient of these was that "production facilities should be located with due regard for strategic considerations," a point especially dear to the American chiefs of staff.[9] Another principle was that the three countries should oppose development of atomic energy in other countries—a position that would have hardly been pleasing to the Belgians, who were contributing most of the uranium and thought the Tripartite Accord assured them eventual participation in atomic energy development.

The outline for a new bargain with the British was emerging. Under the *modus vivendi*, some technical information was exchanged in return for an abrogation of the 50-50 split of raw material supplies from the Congo. Hickenlooper's refusal to allow plutonium to be included in the nine areas of exchange, however, deprived the *modus vivendi* of much of its value to the British. Fear that British disappointment might endanger American access to South African ores prodded the Americans to contemplate a new arrangement; in return for information on plutonium and weapons manufacture, the United States would have a voice in determining the location of British plants, which for strategic reasons the U.S. military wanted in North America. In practice, this would mean that the British effort would keep to a scale which would require only 10 percent of the raw materials available over the next five years. Strategic and fissionable material would be stockpiled in North America; any expansion of British production would be located there, and nuclear weapons, excepting those few stipulated for the United Kingdom by common war plans, would be stored in North America also.[10]

The proposal found support from several members of the AEC, Strauss excepted, and from a special committee of the National Security Council (NSC). This committee consisted of the three American members of the Combined Policy Committee: Acheson, now secretary of state; Secretary of Defense James V. Forrestal; and Sumner T. Pike, acting chairman of the AEC in Lilienthal's absence. On 1 March the staff of the NSC special committee, on which R. Gordon Arneson, now special assistant to the secretary of state for atomic energy, played a key role, forwarded a paper to Acheson. It argued that the exchange of information envisioned, though of greater magnitude than under the *modus vivendi*, still fit the legal interpretation that had allowed the more limited exchange. Section 1(a) of the Atomic Energy Act stated that atomic energy policy was "subject at all times to the paramount objective of assuring the common defense and security."[11] This was the guiding principle which controlled all other sections of the act, including the somewhat more restrictive wording of Section 10(a)1. The staff was sure the exchanges planned were to the benefit of U.S. security. The following day General Dwight D. Eisenhower expressed his view that the new proposal "would restore that mutual trust and confidence among the three nations so essential to the strengthening of our own common defense and security" and volunteered so to testify.[12]

The special committee's formal report to President Harry S Truman rejected the "superficial situation" of continuing the present *modus vivendi*; the difficulties and inefficient utilization of resources and effort were not acceptable. Nor would U.S. pressure force Britain to halt her program work. She would go ahead and also probably withdraw from the European Recovery Program, thus weakening the ERP and producing a victory for Communist forces. Expanding interchange to incorporate all fields of atomic energy including weapons seemed the most sound path. The United Kingdom would gain atomic weapons and plutonium with or without U.S. help. Its influence on Belgium and South Africa should not be underrated. Full collaboration would be consistent with the main lines of U.S. foreign policy and probably enable the

United States to "gain greater initial advantage in the field of allocation of ore."[13] A long-term arrangement of perhaps twenty years would best assure effective cooperation.

The president accepted the proposal for expanded interchange, but the difficulty of gaining congressional support loomed high. McMahon feared a strong reaction from the Joint Committee on Atomic Energy Affairs; the president therefore agreed to meet personally with key legislators, which he did on the steamy evening of 19 July 1949 during a tense gathering at Blair House. Despite Truman's leverage and Eisenhower's support, the meeting did not go well; both McMahon and Hickenlooper questioned the legality of the administration's interpretation of the Atomic Energy Act. When Acheson formally presented the proposal to the Joint Committee on 20 July, he was raked over the coals by Senators Vandenberg, Hickenlooper, and William Knowland. The new secretary of defense, Louis A. Johnson, hastily arranged an adjournment of the meeting. A week later tempers had cooled, in part because Truman let it be known that he did not intend to push the proposal through on the basis of executive privilege. With the understanding that the Joint Committee would be kept well informed, talks with the British were authorized.[14]

Consensus was not solid, however. Questions continued about the legality of the proposal. Secretary of Defense Johnson reportedly no longer supported the American desire for tripartite talks. He felt that the United Kingdom was "finished" and there was no sense in trying to bolster it by financial aid or scientific information. He claimed that even the Canadians expressed doubts of British security. There was question, too, whether the British would accept the restrictions on their program which the Americans envisaged. American Lieutenant General Lauris Norstad had, however, opportunity to meet with British Air Marshal Lord Tedder in July and had seen promising indications of British concessions. Tedder thought that if full cooperation could be achieved, the British would not insist on having a major program. Pride would require that they be allowed to develop one or two bombs, but then they might back off. Plutonium production would have to proceed for a similar reason, but it might be shipped to North America.[15]

By early fall a series of decisions and events had further influenced the Americans to stay with their charted course. Recent financial talks with the British had emphasized the importance of American financial aid in picking up British commitments in the Far East, thus making the British more pliable. The state department had noted meanwhile that Britain's reluctance to enter into arrangements which might derogate from her sovereignty had caused her to be something of a wet blanket regarding the movement toward European integration and unification. The department, desiring to see European unification progress, therefore thought increased effort "should be made to link the United Kingdom more closely to the United States and Canada and to get the United Kingdom to disengage itself as much as possible from Continental European problems."[16] Norstad and Kennan feared that an Anglo-American rift on atomic matters could hinder effective military collaboration. Yet an attempt "to extort a solution" would stiffen British backs and lead to failure. Lilienthal had already argued in July that the current arrangement was creating ill will instead of building good relations.[17] The British meanwhile had been cooperative and even sent highly confidential figures on the cost of the British program to be orally reported to the Joint Committee to allay concern about expenditure of Marshall Fund monies. But their patience was obviously coming to an end. "This is a point at which we cannot tolerate without protest the delays and uncertainties which have surrounded the whole question since last February on the U.S. side."[18]

When the Combined Policy Committee held a negotiation session on 20 September 1949, at least some of those present were aware of the news Truman would make public on the 23rd: the Soviets had exploded their own atomic device some days earlier. This development spurred collaboration. Under Secretary of State James E. Webb broached the topic of collaboration carefully, pointing out the importance of sufficient supplies for the expanded bomb production rate currently being contemplated. As was customary, the matter was referred to a subcommittee which soon experienced difficulty. The British wanted a complete atomic energy program for themselves with full exchange, something the American Joint Congressional Committee would not likely accept. The Canadians favored exchange on all items except weapons, in which they had no interest. America's desire to have all production plants in the United States was unacceptable to the British on principle. The Canadians said it was politically impossible to allow the British to operate in Canada the large industrial facilities involved; besides, the British did not have enough dollars. The British therefore argued that construction of plants in their own country, unpalatable though it was to the Americans, was the only viable solution.[19]

Truman pressed the Joint Committee and others for more cooperation during the weeks that the negotiators referred the conflicting positions to their governments. He even expressed his willingness to go to the people over the heads of the Joint Committee. That became unnecessary as the committee itself, shocked by the Russian achievement and believing that the value of secrecy was diminished in comparison with that of collaboration, seemed more amenable.[20] In November, influenced by reports of American scientists in England, the Americans recognized that two British reactors were too near completion to expect the British to acquiesce to their disassembly; a third might be stopped, and a planned diffusion plant limited to pilot size.

When diplomatic negotiations resumed on 28 November, the proposal submitted by the British made clear their intent to have a full range of facilities and control of some bombs. Presumably this could spell future conflict over raw materials. The American response was to suggest that the British could build whatever weapon parts they wished, as long as the proposed joint integrated effort would not be harmed. Shortly there emerged a concept of a truly integrated program from scientific development through weapons production. Once more the matter had to be referred to London.

Meanwhile time was running out on the uranium allocation agreement. The British therefore proposed an interim arrangement for 1950 which, after some amendment, was accepted by the Americans. It allowed for expanded U.S. production needs of 2,934 tons of oxide to be met, after allowance for other sources of production, primarily from Congo ores. If Congo production exceeded that amount, up to 130 tons could be shipped to Britain. Though Congolese production in 1949 had proved greater than anticipated, the Americans were careful to insist that if the Congo production in 1950 fell short of estimates, then needed ore would be shipped from the British stockpile.[21]

At the end of December the British accepted a truly integrated weapons program leading to a maximum production of bombs over the next three years. They also accepted a limitation on the number of their reactors and an exchange of plutonium for about twenty bombs. The loss of a third reactor was not too great a concession for what might be gained, and the money saved would be substantial. Britain would also receive U-235 from the United States for mixture with her own plutonium for the making of a few bombs, while the bulk of the nuclear cores and triggering devices would be made in the United States.[22]

The English had come a long way in meeting American demands. Yet a problem remained in that they claimed the right to engage in production of fissionable materials as long as such an effort did not interfere with the joint program, and they wished also to explore aspects of atomic power not related to weapons. Though these features and the stockpiling of bombs in England would make it difficult to sell the agreement to the Joint Committee, Arneson and Adrian Fisher of the state department thought they were hard to argue against. Of what use was information exchange on the use of U-235 if the British were not allowed to produce any? Questioning of the proposal was sharp, and the opposition of Secretary of Defense Johnson was rumored.[23]

Whether an agreement could have been put together that was satisfactory to both the British and the Joint Congressional Committee will never be known, for on 2 February 1950 the arrest of Dr. Klaus E. J. Fuchs was announced. Born in Germany, naturalized as a British citizen, Fuchs had been among the first British exchange scientists who arrived in North America in 1943. Intimately involved in the bomb program at Los Alamos, he returned to Britain in 1946 but revisited the United States in 1947. He openly admitted giving atomic energy information to the Soviets and was convicted of espionage on 1 March.

Important secrets apparently had been lost to the Russians. The repercussions of the news of Fuchs' treachery were wide ranging. Secretary of Defense Johnson, scarcely an admirer of the British under any circumstances, abruptly called for direct U.S. negotiations with South Africa on uranium purchases, rather than through the Combined Development Agency. Such a move would have disrupted what collaboration on procurement remained, as well as endangering delicate negotiations. Acheson and the AEC commissioners managed to persuade Johnson to back off, but only on the grounds that there would be a thorough review of tripartite atomic relations. The arrest in May of Harry Gold, in June of David Greenglass, and in August of Julius and Ethel Rosenberg demonstrated that Fuchs was not operating as an individual but as part of a well-organized spy ring.

Congressional concern reached fever pitch, augmented by news of the invasion of South Korea by Communist troops from the north on 25 June 1950. In addition, after months of debate at the highest levels of government, President Truman had in January approved a major effort to create a new weapon several times more powerful than the atomic bomb: a hydrogen bomb operating on the basis of fusion rather than fission. Now development work on the "super" was accelerated in part as a result of knowledge of Fuchs' treachery and the possibility that the Soviets might be farther along this road than the United States.[24] Ironically, the uncovery of Fuchs did more to disrupt the

Anglo-American-Canadian atomic alliance, slow the progress of British production, and increase the determination of the British to have their own program of bomb manufacture than most other steps the Russians could have taken short of war.

Aware of the reaction in the United States, the British members of the Combined Policy Committee proposed in April a joint meeting with security representatives from the three countries to examine security arrangements and presumably establish procedures that would reestablish confidence. The possibility of such talks was favorably received by the AEC, but Acheson poured cold water on the notion that any continuation of the revision of the *modus vivendi* could take place during the remaining months of the congressional session prior to the fall elections. Attlee warned that his country could not delay its program, that perhaps even a third atomic pile might be built, an issue "which might produce difficulties between us."[25] Again, the British saw the villain as the McMahon Act, and thus indirectly the now absent Groves and persons of his pattern of thought.

The British needed fuel, and in June they requested an interim allocation of 404 tons of oxide from the unallocated ores stockpiled in England—the stockpile which had not been depleted as Hickenlooper and other American officials anticipated when the *modus vivendi* was negotiated. It took some while for the Americans to work out their posture on this request, and it is evident that Louis Johnson was unhappy with the willingness of the Atomic Energy Commission to consider it.[26] Delay also resulted from the resignation of David Lilienthal from the chairmanship of that group and the tardy nomination of Gordon E. Dean as his replacement. The invasion of South Korea also had disrupting impact and altered viewpoints.

At a September 1950 CPC meeting, Johnson noted that the joint staffs would soon be submitting a statement that their requirements for atomic weapons had doubled. Could Congo production be increased? Dean replied that negotiations with Sengier were in progress and that perhaps 500 more tons of oxide would be realized in this manner. (Sengier did agree in October to an increase, eased by a price rise to offset the cost of more rapid but less efficient production.)[27] In the end, the British request was granted.

The awkwardness of not having resolved tripartite atomic relations was evident, and the Commission and the Pentagon saw the need of securing plutonium production from the new British reactors. General Omar Bradley, chairman of the joint chiefs of staff, called for all-out cooperation in the weapons field, suggesting that the United States trade bombs for plutonium, though he felt sure the British would wish to produce a few bombs themselves; they had already requested use of the American test site at Eniwetok. All agreed that congressional approval was necessary, that it could not come from the current congress, and that a clear position should be developed in the next two months for discussion with the British. This would in no way conflict with Acheson's 5 October assurance to the Joint Committee that no tripartite accord would be concluded without the knowledge of that committee.[28] Just as the Fuchs betrayal was a *force majeure* working against collaboration, so were events in Korea promoting resumption of exchange.

Amendment of the Atomic Energy Act

Gordon Dean, as new chairman of the AEC, was convinced that most of the Commission's problems of relations with the Joint Congressional Committee lay with the vague wording of Section 10(a) of the McMahon Act. That section allowed the Commission to control data with the purpose of assuring security and common defense, but not to release data applicable to industrial purposes. Not only did this clause frustrate Article 9(a) of the Belgian agreement, but it confused the entire information issue because energy for industrial uses and energy for defense purposes were so closely linked.

As a friend and former law partner of McMahon, Dean thought he could work something out. Unlike Lilienthal, he was not hampered by poor relations with Hickenlooper, who had even held congressional hearings accusing the Atomic Energy Commission and its former chairman of "incredible mismanagement." Moreover, Dean was aware that the expanding uranium hunt would require guidance on relations not only with Britain and Canada under the *modus vivendi* but with other nations as well. South Africa was already asking for a privileged position.

Dean's desire to get matters moving received unexpected assistance at the beginning of 1951, when debate in the House of Commons led Churchill—still in opposition—to ask that the wartime agreement on atomic partnership (that of Quebec) be published, inasmuch as other questioning had revealed that it had been superseded (by the *modus vivendi*). The former prime minister was disturbed especially because his country had abandoned the right to consent or even to consult on

145

use of the bomb by the Americans while leasing airbases in East Anglia for the American planes that would carry those bombs. Other members of parliament also raised sharp questions. The department of state was horrified and quickly had Truman quash the idea of publication because it would create a furor, lead to questioning about current relations, and make any new agreement more difficult to reach.[29]

Under criticism because of the easy advance of North Korean forces in the initial stages of the Korean War, Louis Johnson had resigned as secretary of defense and was succeeded on 21 September 1950 by the aging Marshall, called back from retirement. Marshall's attitudes toward the British and collaboration were basically different from Johnson's, but events in the Far East demanded his full attention; atomic energy affairs were mostly left to deputies.

The decision to produce more bombs of course had its effect. As early as 2 October 1950 Marshall, Dean, and Acheson had minuted the president that

> The proposed expansion program is in accord with, and is limited by the foreseeable supply of uranium ore. In this regard, it is recognized that to facilitate carrying this program forward successfully, the problem of cooperation with the United Kingdom and Canada in the field of atomic energy needs to be resolved.[30]

The joint chiefs of staff also saw the need for cooperation and, by 31 January 1951, had made suggestions for a new overall agreement. Basing their views on a study undertaken by the Military Liaison Committee, they essentially called for an exchange of completed weapons for British plutonium.[31] On 8 May Dean sent Marshall and Acheson a long memorandum outlining his views and requesting a special meeting. Accompanying his memo were reports on the impact of the Atomic Energy Act on scientific intelligence operations and a review of information exchange. He pointed out the need for collective strength, rather than competition, in facing the Soviet threat. The *modus vivendi* was not satisfactory, and ore allocation was still not settled. Tripartite collaboration was needed on declassification of information. Britain could use American U-235 and finished weapons, while America could use British plutonium. This material the Canadians also had and were willing to sell to the United States: moreover, their heavy water reactor at Chalk River could perform high-intensity irradiations which could not be done in the United States.[32]

Dean noted the six objectives desired by the department of defense: (1) exploitation of the best sources of ore in the non-Communist world and the maximizing of supplies for the United States, (2) conversion by the United Kingdom and Canada of their ores to plutonium as quickly as possible, (3) arming of the United Kingdom with nuclear weapons in numbers consistent with the value of British plutonium given to the United States, (4) coordination of the two powers' war capabilities prior to any war, (5) similar maximizing of active and passive defenses, and (6) continuation of existing arrangements regarding Canadian ore and plutonium. The two caveats presented by the defense establishment focused on security of classified information and the location as safely as possible of production facilities, weapons, and delivery vehicles—another wording of the old build-and-store in North America argument.

In his memorandum the AEC chairman described cases of needed cooperation, cited issues with Belgium and South Africa which could not be resolved until Anglo-Canadian-American relations were settled, and proposed negotiation in eight areas ranging from security and raw materials to exchange of technical personnel. He closed by suggesting a substitute for Section 10(a)1 of the McMahon Act:

> That any arrangements entered into after the date of this provision which involves [*sic*] the communication of restricted data to other nations, shall not be placed in effect until the Commission, after having obtained the concurrence of the National Security Council, has determined that the arrangement will promote the security of the United States.[33]

A somewhat similar message by Dean to the same recipients two months later indicates what reaction must have been stimulated, for the amendment was now revised to read that any arrangement for communication of information "shall not be placed in effect until after the Joint Committee on Atomic Energy has been informed and the President has determined that the arrangement will promote the security of the United States."[34] The Joint Committee could not be ignored.

Nor could the Military Liaison Committee, whose chairman, Robert LeBaron, deputy to the secretary of defense for atomic energy affairs, proved difficult. His concern for security of weapons information was so total that he was scarcely willing to allow exchange even on minor bits of information which could definitely aid the United States. At a meeting of the American members of the CPC in August, LeBaron used every opportunity to argue against the need for new legislation, while Dean took the opposite stance. Dean gained some support from Acheson, who suggested that the Joint Committee be asked for its approval of any exchange, rather than just be informed.[35]

LeBaron's concern about security was well founded, for on 7 June 1951 the disappearance of two British foreign service officers was announced. Guy Burgess had been second secretary of the British embassy in Washington from August 1950 to May 1951. Donald D. Maclean had been the recent head of the American division of the foreign office, a first secretary in Washington, and from January 1947 to August 1948 British secretary to the CPC, thus privy to all its debates and documents, including the reports of the Combined Development Trust. The two subsequently turned up in the Soviet Union. Another British defector was Dr. Bruno Pontecorvo, a senior scientist at the British Harwell research center, who had fled in October 1950. Security issues also were rumored in connection with the resignations from the British program of two other leading scientists.[36]

The preoccupation with security of LeBaron and Deputy Secretary of Defense Robert Lovett was especially evident in the discussions of the August gathering of the American CPC members. The chief topic was a British request for shallow water atomic test facilities. At first the department of defense was favorable, on the assumption "that it might be possible to conduct a test in such a manner that the United States would be getting information about the United Kingdom weapon without giving up any United States weapons *quid pro quo*."[37] But if it were impossible to "circumscribe the congenial habits of scientists," then very few British scientists should be allowed, and American scientists could provide the instrument readings. Eniwetok was not available, ostensibly because it was not currently equipped but more likely because American tests, including that of a fusion bomb, were scheduled for the site the next fall.

As discussion proceeded it became clear that whether or not the test was a matter of British pride, as LeBaron thought, a good deal more information had to be interchanged than the soldiers initially thought, even if American scientists were to set off the bomb. The discussion paralleled what was being discovered about bombs that might be exchanged for plutonium: "that it was not practical to give completed weapons to the British without disclosing highly significant technical information on US atomic weapons."[38]

A peculiar stand-off developed. Acheson and the department of state wished to accommodate the British as a means of improving relations. Dean wanted to approve the test but argued that the Commission could not legally do so without amendment of the McMahon Act—too much information would have to be exchanged on weapons, a matter explicitly excluded from exchange under the *modus vivendi*. The department of defense, not wishing any revision of the law that might allow increased information exchange, insisted that the test be so conducted and restricted that it could fit under the existing law; the department's support was contingent on that condition.[39] The British were anxious for a reply, having waited since April. As there seemed insufficient time to push new legislation through, especially against department of defense wishes, that department's position won out. A counterproposal was submitted to the British on 18 September and was found basically acceptable. Some wrinkles had to be ironed out; talks continued through the middle of October, resulting in amplification of the U.S. response in terms of financial, technical, and logistical matters. All seemed agreed: then on 26 December the British rejected the offer in favor of developing their own test site in Australia. They recognized limitations placed by U.S. law. On the other hand, the test they really wanted in order to understand the effect of a bomb blast in English harbors was a shallow-water test, which the United States ruled out because of the unavailability of the Eniwetok site.

> Furthermore, your counter proposals cannot provide for full cooperation and reciprocity in that they prohibit access by our scientists to U.S. restricted data. This is a difficult limitation for us, since in the nature of things we have no say in defining what constitutes "restricted data" or in interpreting the applicability of the definitions in specific cases.[40]

The crux of the issue again seemed to be that of information sharing and the insistence of the Americans on controlling British knowledge. Though the British were gracious about the matter, American policy had forced the British cabinet to take one more significant step toward the development of what the American military did not want, a full-fledged, independent British atomic program. The department of defense had conceded that the British would wish to make and detonate a few of their own bombs; but after the costs of developing the Australian site had been paid, there would be one less factor curbing the size of the British program.

By the time the British decision was reached, two events had occurred in October which, had they taken place in the summer, might have altered the U.S. position significantly. One was the explosion of two test bombs in the Soviet Union, the other the amendment of the McMahon Act.

Both Secretary of State Acheson and Lovett, who succeeded Marshall on 17 September 1951 as secretary of defense, thought it unlikely that congress would tolerate any amendment of the Atomic Energy Act. Dean was not so sure and definitely confident of his rapport with the Joint Committee. He needed a good issue on which to base his demand for a change in the law. The British test request was not quite right, but uranium oxide supply gave him all the leverage he needed. At the 24 August meeting of the American members of the CPC, Dean reported that important ore discoveries had been made in Australia. The cost of shipping the ore would be extreme. The United States could show the Australians how to process and enrich the uranium prior to shipment, but not until the McMahon Act was changed. A decision had to be reached whether 500 tons of uranium per year was worth the risk of making available to the Australians chemical processes that might be leaked to the Russians.[41]

It was Canadian ore that tipped the balance. A large discovery in the Athabaska region suggested that in a few years Canadian production would rival that from the dwindling Congo reserves. The Canadians wanted to hire an American firm to construct a plant to extract the valuable oxides. Canadian processes were less efficient than the newly developed American system. Failure to exchange the necessary information would result in a serious wastage of ore. Canada was also a good opener for discussions; her leaders had long made clear they were interested not in weapon production but in preparation for eventual industrial uses.[42]

General Omar Bradley, chairman of the joint chiefs of staff, had meanwhile eased slightly the hard-line posture of the defense department. A September statement by that department revealed its continuing reluctance to exchange information. Only when the British production of plutonium was truly substantial and able to fuel a "considerable number" of weapons, or when there was "a better chance that the defense of Western Europe will be effective," could the department alter its stance. The department's preference for a rearrangement of the handling of all atomic weapons matters through military channels was also evident. On this point the current AEC was as unwilling to yield as it had been under Lilienthal's chairmanship. But, the defense department's new statement suggested, if the Joint Committee would strongly recommend an amendment that allowed exchange deemed by the president to be in the nation's interest and "place in the Department of Defense the safeguarding of weapons information," the department would support the legislation.[43]

The bargain was clear. The AEC would relinquish its authority to control weapons information exchange, and the amendment could proceed. The AEC sacrifice would not be great, for the commission would continue to control most matters and the Military Liaison Committee already had a fairly effective voice on information exchange. Leakage of sensitive atomic information, the defense department held, "is the most damaging single factor in the present determination of U.S. military posture."[44] Case-by-case decisions by the president were better than broad charters, and the department had no desire to restrict the AEC on matters of policy.

Dean took his case to the Joint Committee on 12 September, where Hickenlooper was the key figure. When he came to see that some arrangement applicable to more countries than just Canada was necessary, the focus shifted to careful wording. Meanwhile Lovett formally retracted the department of defense position of 31 January agreeing to cooperation with the United Kingdom and Canada on atomic weapons on the grounds that he had learned that an exchange of weapons for plutonium would require the exchange of too many technical secrets. Dean strove on, apparently with no help from Acheson or Lovett, substantial support from Bradley, and reported determined resistance from LeBaron.[45]

In the end, with Truman's approval, the Joint Congressional Committee fashioned an amendment that permitted any specific AEC proposal for information exchange, excluding weapon information, just as had the *modus vivendi* but now not limited to the nine points of that agreement. A proposal would go first to the National Security Council (where the secretary of defense had a voice) and then to the president. Moreover, the proposal would have to lie before the Joint Committee for thirty days while congress was in session.

The Joint Committee's unanimous approval assured passage, and the bill amending the Atomic Energy Act became law on 30 October

1951. Some in the department of defense resented this success. On 2 November Acting Secretary William C. Foster wrote Dean reproaching him for failure to consult the department sufficiently or to allow the joint chiefs or the Military Liaison Committee to have enough input. Dean pointed out that he had kept Lovett informed and gave him a copy of the legislation eight days prior to its being reported out of committee.[46] Perhaps Foster was misinformed, or perhaps Lovett had purposely let the matter ride. In any case, LeBaron apparently had little chance and time to build opposition. Though the joint chiefs had really not given up anything more than they had conceded in early September, they had not received the *quid pro quo* of acknowledged control of all atomic weapons information, an important step to the total control of atomic weapons they desired. An important log jam had been broken, one that might not have opened for many more months save for the uranium shortage. The Korean War, the Soviet atomic tests, and discoveries of Soviet spies increased the American military's anxiety, enhancing its desire both for secrecy and for weapons. It could not have its cake and eat it too. As for the AEC, it soon moved to gain approval of the transmission of the necessary processing technology to Canada.[47]

Faltering Communication

Passage of the legislation had not, however, reconciled the difference of views between the AEC and some members of the department of defense, differences which ranged well beyond the issue of information exchange to weapon control. The controversy reappeared when Winston Churchill, prime minister following the Conservative party electoral victory of October 1951, came to Washington in January 1952. One of his key purposes was to get exchange moving. The department of state was prepared to consider this in the existing *modus vivendi*'s areas but wanted to know if British progress meant that now there was more to be learned from their programs. The joint chiefs wished that the U.S. freedom to use the bomb on its own decision be clearly understood and not hindered in any way. (Although the Quebec Agreement requiring British consent was defunct, the *modus vivendi* still called for prior consultation.)

At a 7 January meeting with Truman, Churchill pressed for renewed exchange under the *modus vivendi* and assured Acheson that Britain's new security measures would be adequate. No advocate of bulk production of bombs in Britain, Churchill nevertheless was determined that his nation should have state-of-the-art technical knowledge— knowledge he considered Britain's just due for scientific contributions now several years in the past—and be spared unnecessary expenses. At the White House the atmosphere was cooperative, but that was hardly the case on the tenth when Dean, LeBaron, and Arneson met with Churchill's scientific advisor, Lord Cherwell. The notion of exchange was admitted in principle, but when Cherwell identified specific areas, Dean worried whether they verged on weapon information. LeBaron was more outspoken. He claimed the defense department had no voice in the phrasing of the amendment and felt free to judge each request for information as it saw fit. When Cherwell retorted that such an attitude could lead only to meaningless exchange, LeBaron asserted that all significant atomic energy data fell into the area of weapons.[48] A stand-off had appeared once again.

This time, as during the war years, the heads of government had plainly expressed the desire that information exchange be renewed. Arneson quickly memoed Secretary of State Acheson that "it seems clear that the spirit and intent of the Truman-Churchill exchange has not been instilled in Mr. LeBaron."[49] He asked that the department of defense be reminded of the president's views. Fortunately for Anglo-American relations, Cherwell's meeting later the same day with General Walter Bedell Smith, head of the American Central Intelligence Agency, produced more positive results. Smith favored wider exchange and agreed to receive specific requests for information which would be handled according to the amended Atomic Energy Act rules. Renewed exchange would be tried for a year to see just what could be worked out. Though such arrangement forecast occasional squalls, still important steps might be taken as with Canada. Moreover, the continuing exchange would provide a more positive basis for the yearly discussion of ore allocation.

The results of the meeting were especially disappointing to the British in that some of them had earlier gained the impression that Truman's assent to fuller cooperation meant maximum cooperation within the McMahon Act, and they apparently took a broader interpretation of the law than did the Americans.[50] If the British and American records of the conversation of the two heads of government differed on the point of the extent of cooperation envisioned, it was not the first time that mis-communication on this issue had occurred. The British, it seems, felt so justified in their position that they may occasionally have heard what they thought the Americans *should* be saying. Truman did not like to get pinned down to niceties and, with traditional Missouri pragmatism, probably intended to encourage the notion of seeing what could be done. The legalism of the British—and the accounts of information exchange from their side make frequent reference to "fairness," "equity," and "justice"—led them to believe that such openness joined with the high principles the British saw involved spelled a great deal of cooperation. This hope ran headlong into legalistically narrow interpretations of the McMahon Act and the more important point that some Americans simply saw little utility in sharing and some danger, in terms both of leaks to the Soviets and of criticism by congress. Whether they were discussing generalities or legal obligations, the representatives of the two countries were conversing at different and non-congruous levels of abstraction and communication.

The rapidly growing lead of the United States in production techniques which persuaded the British to continue to call for exchange despite humiliating rebuffs was the very factor which caused men like LeBaron to think that exchange would be of little pragmatic value to the Americans. To dispel that impression, the British were generous in sharing information; in return, they received little information but a concomitant augmentation of their bitterness. Cherwell's arguments about avoiding waste of funds and effort were valid. Yet by this time the Americans' research thrust was on development of a thermonuclear bomb, and their production effort was geared to the manufacture of twenty-five atomic bombs per working day; the joint chiefs of staff and the special committee of the NSC were recommending a further 50 percent augmentation of plutonium production and a 150 percent increase of ore alloy production.[51] Even if the relatively small British experimental program did come up with a new idea, how easily could that idea be incorporated into the massive and expensive production facilities already in operation and growing in the United States?

British efforts to continue exchange under the cumbersome provisions of the McMahon Act which the Americans insisted on following for each particular request thus did not yield much satisfaction throughout 1952. The Londoners hoped that their own bomb test in the early fall might cause Washington to show greater interest and respect for what British scientists were achieving. The test at Monte Bello, Australia, the third of October was a success; but its flash of achievement was soon obliterated by the even greater burst of the first hydrogen bomb at the U.S. trials at Eniwetok the beginning of November. Despite the British achievement, the gap between the two nations' nuclear programs seemed greater than ever, and the unwillingness of the United States to enter into serious exchange appeared all the more like the arrogance of a great power reminding another power of its lower status. As the British ambassador to Washington, Sir Oliver Franks, had written in December 1948,

> the whole question of our relations with the Americans on atomic energy questions seems to me to be becoming increasingly bound up with the larger issue of the extent to which the Americans are prepared to treat us on more or less equal terms as a first-class power.[52]

Nothing had transpired since then to dispell this impression, and the very collaboration that did begin to build between Britain and the United States in matters other than atomic only put this issue in greater relief. It is no wonder that Churchill, of all people, did not hesitate to renew pressure on the Americans for better collaboration.

If the prime minister's January 1952 discussion with Truman on information exchange did not produce significant movement in the ensuing months, Churchill was able to nudge along another nuclear-related matter significant both to the U.K.'s position in the alliance and to her sovereignty.

During the negotiation of the hapless *modus vivendi* of January 1948, Clause II of the Quebec Agreement requiring consent of both signatory powers to use of the bomb had been discarded as a counter for the abandonment of Clause IV, which restricted British commercial development of the atom. Churchill had tried to embarrass the Attlee government on the matter to no avail because the Americans refused to reveal the terms of the *modus vivendi*.

At the time of the Berlin blockade crisis of 1948, Britain had granted a hasty American request that U.S. bombers be stationed at British airfields. Years later the British asserted that the American ambassador had given oral assurances that atomic bombs would not be used without prior consultation; no evidence of such assurances was available, however. The foreign office in any case remained concerned on this point. Following the outbreak of the Korean War, Attlee raised the matter with Truman. According to British sources, Truman acknowledged the bomb as a joint possession of the United Kingdom, the United States, and Canada, not to be used save in an extreme emergency such as a sudden attack on the United States. Dean Acheson relates in his memoirs that, following a private talk with Attlee, Truman admitted making some such comment. Lovett and Acheson, aware of the president's many formal commitments not to limit his power to authorize use of the bomb, reacted quickly. Such limitation was not allowed by U.S. law, nor would congress admit it. They forced a quick conference at which Attlee "a little sadly" acknowledged this circumstance and agreed to an appropriately bland statement.[53]

Though no lasting and formal commitment was made—nor could legally be made by the president, as the British well knew—some Englishmen either ignored the retraction so promptly arranged by Acheson or asserted that the president had broken his word. Once again there was mis-communication, augmented by the looseness of the American style and the eagerness of the British to attribute binding specifics to friendly generalities.

The problem of the bombers remained, provoking further negotiations. The Britishers' sovereignty was involved, and they rightly had no desire for retaliatory attacks to be directed to their bases as a result of U.S. bombing of Russia from either there or elsewhere. American diplomats recognized the legitimacy of the concern; while they would not make concessions regarding bases not located in the United States, in October 1951 they agreed to a statement acknowledging that use of U.K. bases and facilities was by nature a matter for joint discussions. Sir Oliver Franks' careful wording avoided raising the hackles of congress while assuring prior consultation.

In January 1952 Churchill returned to the charge while in Washington. He achieved a published assurance that, in the event of war, British bases would not be used without British assent. Yet the real probability remained that those bases and the rest of the country might be bombed in retaliation for an attack launched elsewhere. The United States would give no further assurances but did take steps to share its air strategy and planning with the British in subsequent months.[54] Thus did the search for the bomb and the bargaining of scientific information for uranium lead logically and perhaps inevitably to closer collaboration in higher strategy, the very matter which prompted the original Quebec accord.

Despite the lack of agreement on atomic energy matters, general relations between Great Britain and the United States had steadily improved since the middle of 1947. Cooperation had been close during the war, and it was to be expected that some falling apart would occur in the years following, as each power addressed its own domestic and international needs. But the Marshall Plan, the Communist coup in Czechoslovakia, the formation of NATO, the explosion of the Russian atomic bomb, the United Nations effort in Korea, all had their inter-related effects in pulling the United States and Britain together. Defense and non-nuclear armaments were mutually discussed. Atomic energy was the only major aberration in the pattern. Even when information exchange was at its nadir, agreement was reached on uranium allocation. The need for a sufficient supply of the ore, joined with the production of scores of bombs and the consequent shift of the security emphasis from bomb manufacture to bomb delivery, would militate toward a more flexible American position. The British could also hope that the American political campaign for a successor to Truman, who had chosen not to run for the presidency in 1952, might bring some alteration of the existing stalemate. The election of Dwight D. Eisen-hower, long known for his support of allied cooperation, buoyed their spirits.

7 · *Much Effort, Limited Gain: Continuing Global Negotiations*

The debates and divisions which arose over information sharing among the closest of allies at the end of the war signalled the possibility that other wartime arrangements might come unstuck. Careful negotiation between the British and Americans was required simply to maintain the spirit and principle of the preemption accords. The ability of those accords to provide needed materials quickly proved doubtful. Yet new potential sources kept appearing which in turn required examination and negotiation. At the same time, increasing numbers of countries were launching atomic research programs and requesting information, raw materials, and expensive equipment. By the end of the decade allied attempts to cover all these matters as well as to deal with the many other serious issues of the day began to resemble those of a cook with too many pans on the fire.

The expansion of the search for uranium ores was stimulated by the continuing deterioration of Western relations with the Soviet Union and the awareness that the only effective counter the West held to the masses of Russian conventional troops and armor was an overwhelming preponderance of nuclear arms and the will to use them. Still in the future lay the concept of nuclear forces being relatively balanced between the super powers so as to provide a stabilized condition conducive to peace and avoidance of "mutually assured destruction." The Redox system, which would achieve far more efficiency in the processing of uranium and in the use of low-grade ores, was not yet perfected. The discovery of pitchblende in southern Utah, which would change the course and fortunes of the uranium search in the United States, was yet a year away. Though the United States Atomic Energy Commission took the more aggressive role, the British remained significant joint partners in the effort. It was they who took the lead in Portugal and India, while the Americans played the opening cards in Brazil and Spain. Although the Americans sometimes chafed at what they considered British softness in terms of willingness to give information and technical assistance to the Norwegians and French, they had to acknowledge the long-standing British diplomatic influence in Belgium and Scandinavia and its legacy of experience, knowledge, and trust which the United States could not match. Too, the extensive territories of the Commonwealth still held promise.

Reworking the Brazilian Accord

The situation in Brazil was complicated by domestic events in that country. Vargas had repeatedly promised free elections in which he would not be a candidate once the war was over; now he was confronted with the need to carry out his statements. Censorship was lifted in February 1946, and eventually even the previously banned Communist party was recognized. Vargas threw his personal support to his minister of war, Eurico Gaspar Dutra, as candidate for president, urging the electorate to support their Brazilian Labor party. The National Democratic Union backed Brigadier Eduardo Gomes. The military was split. Some feared that Vargas might attempt a coup or postpone the elections. To forestall such possibilities, a group of officers staged a bloodless coup in October 1946, forcing Vargas to retire and turning over the government leadership to his constitutional successor, the chief justice. Elections were held in December, with the Labor party and Vargas' candidate winning handsomely. In the campaign, however, Dutra committed himself to support of a constituent assembly and a revision of Vargas' constitution of 1937. A more liberal constitution was promulgated 18 September 1946. Its terms, the watchful eye of the democratic and Communist opposition, and the now unfettered press all meant that implementation of the 1945 agreement with the United States on thorium faced difficulties.

Thus it was that, although Groves and the Combined Policy Committee agreed in December of 1945 that it was time to purchase material from Brazil, little was accomplished. Negotiations were begun in conjunction with various American firms interested more in components other than thorium in the monazite sands, yet over the next two years the yield was only about 1,000 tons of sands annually. Despite the provision in the 1945 accord that the United States purchase a minimum of 3,000 tons per year, the lesser amount was all the Brazilians would make available.[1]

Meanwhile word reached Washington of offers made by French, Dutch, Portuguese, Canadian, and other firms to purchase Brazilian monazite. The Canadians were especially interested in obtaining thorium for research purposes. The matter was discussed at a 3 February 1947 Combined Policy Committee meeting, where their request for 75 tons of monazite sands was judged reasonable, referred to the American Atomic Energy Commission, and hopes expressed that shipments from Travancore in India would soon be available. The Vargas agreement called for prior American approval of Brazilian shipment of monazite or thorium to any third parties. Therefore the Americans were slightly annoyed when they learned from the Canadians in March 1947 that Brazil had offered to sell Canada a ton of the material, with an exchange of scientists being part of the deal. Though the Brazilians later suggested that the Canadians initiated the discussions, the state department learned from its own sources that in fact the opposite had occurred.[2]

The amount involved was small, and the United States had no objection to Canada's obtaining the sands, especially as the Canadians rejected any scientific exchange. Yet the episode and the paucity of shipments warned that Brazil was not deeply committed to the 1945 accord. Moreover, inspired editorials were appearing in Rio newspapers advocating nationalization of monazite deposits and trade, with higher price and scientific concessions from monazite purchasers, the chief of which was acknowledged to be the United States. The Lindsay Light and Chemical Company, an American firm which processed monazite sands, was being asked to build refining plants in Brazil.

Edmund Gullion and Under Secretary of State Acheson held in March 1947 that the U.S. attitude toward Brazilian nationalization depended on how it would affect the 1945 agreement, American access to material, and price; any arrangement which would allow thorium residues to remain in Brazil they would also oppose. The rising suspicion that Brazil would be asking for renegotiation of the 1945 agreement prior to its first renewal in 1948 was confirmed in July 1947 by President Dutra to the American ambassador, William Pawley. The ambassador, for his part, welcomed the chance for government-to-government negotiations, as small-scale commercial dealings had been ineffective. A problem loomed in that Pawley was fairly sure that the Brazilians would insist that any agreement be ratified by the Brazilian senate, while the United States would insist on a secret executive agreement. Then, too, there was the matter of price. This was supposed to be renegotiated every three years. In his preliminary approach, President Dutra and his cabinet members had suggested a figure of $75 per ton, a hefty increase from the 1945 settlement, which called for a grad-

Brazil & India main — or only hi-grade sources of thorium

ed price range of $22 to $40 per metric ton according to thoria content. The Brazilians would also likely require some processing facilities in their country. Finally, they had been making comments in the UN to the effect they would reserve the right to decide what amount of their resources they would make available to any international authority, thus indicating a possible reluctance to allow the United States to exhaust Brazil's monazite reserves under the terms of the 1945 agreement.[3]

The Americans were not surprised by the Brazilian wish to negotiate and were willing to make price concessions and other adjustments. At the same time they knew they held some counters in terms of various credit and aid programs which could be used to exert pressure. Their opening effort would be to remind the Brazilians that they did not have the only source of monazite, that common security called for close collaboration, and that the issue was a major point in Brazilian-American relations. Aware of French inquiries in Rio, the state department informally called to the Brazilians' attention that any commitment made to the French would be contrary to the 1945 agreement which the United States considered as remaining in effect until modified by mutual consent. This last was an important point, both because the department did not want the materials given to third parties and because negotiation of any new agreement, rather than amendment of the old, might give the Brazilian National Security Council opportunity to confuse the issue or press for an arrangement less advantageous to the United States.[4]

Offers from Washington to talk about prices and quantities produced no response; the expiration date of the first term of the contract approached without confirmation of the Americans' declaration of intent to renew. When in April 1948 Brazilian Foreign Minister Raul Fernandes was jogged into a reply, the Americans got what they did not want to hear. Fernandes thought the 1945 agreement illegal and unenforceable. Vargas had exceeded even his own extended powers in making an executive agreement which received no touch of legality through any form of publication in an official government organ. Moreover, without specific legislation passed by the Brazilian congress, the current constitutional government could not force exporters to ship to any particular country. A new agreement, also advocated by the Brazilian National Security Council, should be written which would reserve part of production for Brazil, call for maximum treatment of the sands there before exportation, and contain an escape clause which would allow Brazil to adhere to any plan adopted by the United Nations. The government would also take steps to expropriate all monazite mining concessions in order to assure national control.[5]

Mining of monazite was not the only area in which Brazilian resistance to American influence was becoming evident. In the opening months of 1948 the Brazilian congress, able to assert itself more fully than under Vargas and reflecting growing nationalism, was in the process of passing legislation limiting foreign investment in petroleum exploration, refining, and transportation only to production for exportation; the domestic market was to be left solely to Brazilian-controlled firms. American oil companies were upset. The state department recognized that any government association with their complaints could be counter-productive, yet it discreetly became involved in arguing the case for more extensive involvement of foreign capital without legislative restraints. During the last months of the year, a special economic mission visited Brazil in an attempt to work out inter-governmental arrangements on the role of foreign capital in Brazilian development. To some members of the petroleum division within the department of state it seemed that the best chance for a favorable petroleum law and for supplies of diamonds, quartz crystal, and monazite sands lay in a trade by which the United States would agree to help Brazil, financially and technically, in various development projects.[6]

Brazil was indeed in need of considerable financial aid. After much negotiation the Federal Reserve Bank in New York City took over a $80,000,000 monetary stabilization loan due the U.S. treasury in July 1948, effectively granting an extension. Other loans for various industrial and military projects were worked out, including a $7.5 million Export-Import Bank advance. In February at state department urging the administrators of the European Recovery Program ruled that 275,000 metric tons of coffee could be imported into Europe during the first fifteen months of the program, thus greatly bolstering Brazilian trade with Marshall Plan dollars.[7]

All of these factors must be kept in mind in considering the shifts taken by Brazilian diplomats in the monazite affair. In June 1948 Rio officials gave formal written response to American inquiries regarding continuation of the 1945 agreement. The proposed new draft agreement was less than acceptable to Washington, primarily because it would permit shipment of monazite outside the Western hemisphere. The Americans did acknowledge a conflict between their desires for secrecy and the Brazilian constitution and promised to study the matter, meanwhile insisting that the old agreement stay in effect. Continuation of that agreement required no publicity and corresponded precisely with American objectives. The United States was therefore content to let the matter ride when Fernandes told Herschel Johnson, now ambassador in Rio rather than in Stockholm, that his assistant had misunderstood his directions in drafting the proposed new treaty and that there was no intent to permit exportation of the sands to other foreign powers. Fernandes also promised to find some way to handle the agreement secretly under Brazilian law.[8]

Yet over the next months few imports of monazite were received in the United States. Existing commercial contracts were due to expire at the beginning of 1950, and rumors mounted of Brazilian intentions to place an embargo on the exportation of monazite, thorium, uranium, and beryllium to any foreign country.[9] Anxious to scotch any movement toward embargo, in December 1949 the United States pressed for a formal statement that the *status quo* regarding monazite would be continued. Again, little happened. The following March Johnson returned to the charge by approaching President Dutra, who concurred that the attacks mounting in the left-wing and nationalist press against the United States and the supposedly secret agreement were unfortunate. Dutra said he had taken steps to quash a secret session of the congress intended to discuss the embargo and the secret accord. To Johnson's relief he added that he was trying to expedite the Brazilian Security Council's plan to set up a government monopoly of fissionable materials under its jurisdiction. The United States, earlier desirous of keeping the matter out of the Council's hands, now was eager for it to gain control.[10]

Progress toward a settlement remained minimal. The British, unable for several years to receive monazite sands from Travancore and needing 250 tons for each of two years for research purposes, hastened to ask for that much from the small amounts still being received by the United States from Brazil. They believed the projected embargo would soon be passed, and they wished the export contracts signed before the law went into effect.[11]

Brazilian elections prevented any legislative action prior to the assembling of the new congress in January of 1950. Meanwhile the United States actively supported Brazilian efforts at international tariff negotiations to reduce import duties on thorium and cerium; such action was seen as enhancing U.S. efforts to obtain monazite. Some Brazilians were anxious that their country profit more directly from its rare reserves, both in its own use of them and in processing the sands for others, but state department and Atomic Energy Commission officials had little interest in conveying the technology or in funding the construction of a processing plant. Ambassador Johnson was therefore relieved when, at a March 1959 meeting, the new Brazilian foreign minister, Neves da Fontoura, seemed to drop his previous pattern of using the processing plant as a direct *quid pro quo* for any monazite agreement. Fontoura did assure continuing supplies of sands for the United States, and Johnson promised to explore the matter of a processing plant. This he did, reporting to his superiors that both material and political interests in Brazil desired the plant, which would probably operate under American control; these interests were "strong and articulate."[12] Among them was Dutra's successor as president, Getulio Vargas. Both Johnson and Gordon Arneson, special assistant to the secretary of state for atomic energy matters, were of the view that the best route would be to interest a private American firm in the project. Any arrangement worked out with a private firm without American government involvement would, of course, help the United States to avoid any embarrassing conflict with other nations, such as Belgium, which were pressing for scientific assistance. But no American firm would take the plunge.

The actual value of constructing a monazite processing plant in Brazil in the early 1950s was doubtful. While some of the other products of the sands were in demand, no atomic energy use of thorium had yet been developed. Indeed, though the Americans' preemptive interest remained, partially processed sands containing several hundred tons of thorium also remained untouched at an American processing plant. The lack of a market for thorium made further refining wasteful. Ultimately, it was the French rather than the Americans who agreed to build the plant.[13]

But there were other nuclear installations that interested the Brazilians. On a visit to Washington the chairman of the Brazilian National Security Council indicated that a cyclotron was on his shopping list. He was assured by diplomatic and AEC officials that the General Electric Corporation would readily be granted an export license to ship a cyclotron. The General Electric executives, however, were far from eager to take the contract. A cyclotron was difficult to build and required materials difficult to obtain. Only at a formal request by the U.S. government would the company move ahead.[14]

The question of putting pressure on General Electric turned, in the opinion of the Atomic Energy Commissioners, on the Brazilian reaction to a special mission by the AEC chairman, Gordon Dean, in the fall of 1951. Now aware of the possibility that uranium could occur in conjunction with thorium deposits, and stimulated by the Soviet achievement and the subsequent American decision markedly to expand bomb production, the Commission was extending its uranium search for procurement as well as preemption reasons. The reappearance of Vargas in the Brazilian presidency and the resolution of some of the arguments over information exchange in Washington suggested that positive results might be achieved from an approach to the Brazilians. Though Dean's visit was primarily associated with the Commission's desire to explore Brazil's uranium reserves, a breakthrough on this would have impact on the monazite issue as well.

As with the negotiation of the 1945 accord, the Brazilian foreign minister was not much involved at first, although preliminary talks with Fontoura in April had gone well. Basic agreement on uranium and general cooperation in the area of atomic energy was already established at the staff level when Fontoura suddenly reentered the picture. The foreign minister announced he had worked out an alternate draft agreement, acceptable to Vargas, which he thought would avoid the political difficulties which might arise if Brazil promised to the United States all her production in excess of her own needs. Instead, he proposed that Brazil make half of all her mined uranium available to the United States. Dean flatly refused. No agreement would be better than such a 50 percent deal, he said; were it to become known to the Belgians, South Africans, and other uranium suppliers, the effect would be highly unfortunate. Fontoura backpeddled, and Dean quickly drafted an alternate to the original proposal worked out by lower-level officials. This committed Brazil to providing 50 percent of all mined uranium ore plus a statement of willingness to let the Americans have that portion of the remaining ore not needed by Brazil in her program. Fontoura acquiesced, if not happily, and Johnson estimated that Vargas' approval would be forthcoming. To be sure, Johnson urged that the U.S. government request General Electric to accept the Brazilian cyclotron order. The Brazilians would otherwise look to the Netherlands firm of Phillips and would doubt the sincerity of United States claims for close cooperation with Brazil in the atomic energy field. Satisfactory conclusion of the uranium agreement could be jeopardized, and in any case its implementation remained dependent on Brazilian good will.[15]

So matters stood at the end of 1951. Agreement seemed within reach. Yet the flavor of the transaction remained different from those with the Netherlands, Belgium, or even Sweden. Though the United States diplomats kept harping on the issue of mutual hemisphere security, the pivotal point was not common concern to reach a mutually beneficial goal, but rather one of commercial bargaining. There is nothing wrong with the latter, and it has been conducted among even the best of allies. In this instance, however, little attention seemed to be given to the notion of shared responsibility. The Brazilian officials were not so much concerned about a Soviet attack on the United States as on what effect such an attack would have on American support for Brazil against the growing ambitions of Juan Peron's Argentina. Conditioned by years of dollar diplomacy, Brazilian—and U.S.—diplomats tended to think primarily in terms of commercial gains. Issues of principle, such as national pride in control of resources or constitutional limits on secret accords, could be set aside, not on broader principles of national responsibility for hemisphere security but on a narrower interpretation of national interest consisting of the best commercial deal possible. The United States recognized this Brazilian tendency and took advantage of it. In so doing, the spokesmen for an open, democratic, and constitutional society once again found themselves working with a past violator of constitutions and hoping that open, democratic, constitutional procedures would be bypassed. In one sense, this course was no different from that followed with European countries. In another, there was an important difference, for in the latter cases constitutional democratic government was well established and agreements, if achieved, were not bought but reached out of mutual agreement on common political responsibilities and goals.

Negotiations with India

The importance of an agreement with Brazil, or at least the avoidance of an embargo on monazite exports, was accentuated by the demise of such trade with Travancore. Negotiations between the British ministry of supply and that state resulted in early 1947 in an agreement for purchase by the British of 9,000 tons of monazite over a three-year period beginning 1 January 1947. In return, the British agreed to use their good offices in Travancore's efforts to persuade a British firm, Thorium Limited, to build a processing plant in the Indian state. To assure that the nascent plant would have a market for its product, the British also agreed to purchase all the thorium nitrate it would produce over the next five years.[16]

The Indian interest in a production plant was tied to aspirations for atomic energy development which the office of U.S. Under Secretary of State Robert Lovett termed "illimitable."[17] Suffering from lack of national fuel supplies and faced with an immense task in modernizing and industrializing the overpopulated subcontinent, Indian officials were naturally interested in the potential industrial prospects of atomic energy. There was little awareness at the time of the extent of research still necessary before thorium might be found a practical fuel for useful industrial atomic energy production. The establishment of processing plants in India would of course enhance the country's foreign exchange position. Moreover, the subcontinent was one of the chief users of incandescent gas mantles, for which thorium nitrate was an essential ingredient. Obviously there was commercial advantage to producing the thorium nitrate in India and providing it to the some thirty mantle firms there rather than shipping monazite for processing all the way to the United States. Finally, there was continuing demand for the cerium also found in monazite deposits, as it was important in the manufacture of high-grade abrasives and illuminating arcs for motion picture projectors and searchlights.

Unfortunately, the task of building a thorium processing plant in Travancore was no more appealing to British industrialists than constructing one in Brazil was to the American, despite the market for gas mantles. Whether delays were the result of foot-dragging on the part of the British, as the Travancore government alleged, or of roadblocks and unreasonable expectations raised by the Indian state is not clear. In any case, by the latter half of 1947 little progress had been made toward the plant's establishment. In September of that year of Indian independence and separation from the British Empire, the Travancore government ceased to grant licenses for the export of thorium to Great Britain. Its former exports of some 3,000 tons per year to the United States, the United Kingdom, Germany, and France dwindled to nothing.

It was for the British to take the lead in discussions with Commonwealth nations as it was for the Americans to play the primary role in dealing with Brazil, although the two great powers were both to participate in the receipt of materials from either agreement. The American position in the Indian situation was a bit different from that of England in the dealings with Brazil. First, this was because several U.S. firms were experiencing shortages of cerium. Unaware of the broader responsibilities of the Combined Development Trust, these firms were agitating for direct American bargaining with Travancore. Such American interference would, of course, have upset the already fragile agreement on resource sharing within the Combined Policy Committee. Second, India was dependent on the United States for processed thorium for its gas mantles. A U.S. embargo on export of thorium nitrate would in effect turn off the lights over much of India. Finally, there was the matter of beryl.

Beryl is a relatively scarce ore, the sole source of the metal beryllium which in an alloy of about 4 percent with copper is used for many specialty purposes because of the non-corrosive, non-sparking, and non-fatiguing properties of the alloy. Scientists had fairly recently discovered that beryl might also serve as a moderator in a nuclear pile to slow the reaction. As such, it might have both structural and financial advantages over the expensive pure graphite or heavy water which were currently in use as moderators. Like thorium and uranium, beryl was therefore an important material in the nuclear arms race. Unlike the first two, it was not subject to the original Quebec Agreement or the arrangements of the Combined Development Trust. The United States could negotiate for it without reference to the British.

Under American stimulation, beryl production in India rose to as much as 1,000 tons per year during the war, but declined sharply thereafter. In 1946 it was placed on an embargo list by Indian officials, and the United States was left to rely upon an insufficient supply from Brazil. Negotiations eventually revealed that individuals in the Indian government hoped for a variety of *quid pro quo*'s in return for release of beryl. These ranged from half a million tons of steel, phosphate rock, or 1 percent of the world's petroleum output, to a coal liquefaction plant or a beryllia processing plant.

Only the last item made sense to the U.S. state department, and it was vetoed by the Atomic Energy Commission, which did not desire the spread of such plants for security reasons: the technique of producing beryllium oxide was known in the free world by only three small American firms. Two Indian scientific institutions, however, were anxious for particle accelerators for atomic research purposes. In May of 1948 the exchange of such equipment for shipments of beryl was the subject of a favorable discussion between the U.S. minerals attaché in India and Dr. Sir Shanti Swarup Bhatnagar, secretary of the Indian department of scientific research and Prime Minister Pandet Jawaharlal Nehru's chief atomic energy adviser. Bhatnagar's influence on Nehru, however, was thought to be waning as it was countered by that of an anti-American Professor Saha of Calcutta, who was feared by U.S. officials to be pro-Soviet.[18]

In addition to his other posts, Bhatnagar was a member of the Indian Atomic Energy Commission created in August 1948 as a result of the Atomic Energy Bill passed by the Indian parliament the preceding April. Among other features of that bill, the exploitation, processing, and export and import of beryllium and radioactive substances were placed under the control of the Indian government.

A further result of the Indian Atomic Energy Act was the transfer of sovereignty in such matters from the state of Travancore to the central government. The British therefore pursued their monazite discussions with the central government and by June 1948 had reached agreement over the size of the monazite processing plant that should be built. Meanwhile, the Americans took steps to achieve their goals of relaxing of the embargoes effected by the Indian government; the increase of stockpiles in the United States of cerium, thorium nitrate, and beryl; and the winning of a promise from the New Delhi government that access to these materials would be denied to unfriendly nations. The apple cart nearly tipped when the leading U.S. exporter of thorium nitrate threatened to suspend shipments to India unless India resumed shipments of monazite from which the firm could acquire cerium. The state department dissuaded the firm, presumably the Lindsay Light and Chemical Company, on the grounds that the ill will such action would arouse in India would scuttle the beryl negotiations, while continued receipt of American supplies by Indian consumers would enhance American interests.

At this point the British concluded that Indian monazite was not worth pressing a British firm to construct a 3,000 ton-per-year processing plant in India, the key condition for resumption of monazite shipping by India. Thorium was simply not likely to be important in the production of atomic energy for some while. Moreover, should it become so, the amount required in relation to the large known supplies meant that preemptive purchases would not be effective. Thorium Limited, the British firm which had reluctantly agreed to build a 1,500-ton-per-year plant, wanted to be relieved of that commitment and had no interest in building a plant twice that size. If the Indians could work out an agreement with another firm, that was their prerogative. The British estimated their own research needs for thorium as no more than 40 tons of thorium nitrate over five years and 100 tons over ten years. The United Kingdom already had 50 tons of oxide on hand, and Lindsay Light and Chemical in the United States had a surplus of 5 tons a month over American needs which it was willing to sell. American needs were not thought to be any greater than the British. Thus the Brazilian supply of 1,500 tons of monazite per year, producing about 90 to 100 tons of THO_2, would be sufficient.[19]

The British position did nothing, of course, to meet the industrial needs for cerium in the United States. Nor did it facilitate the acquisition of beryl. When a general meeting of British and American officials was held in Washington in late October 1948, the Americans requested their partners to prolong the monazite discussions. They wished to preempt those supplies and to give time for a meeting with Charles R. Lindsay, III, to determine if his firm would build a monazite processing plant for the Indians.[20]

As events turned out, it was not the Lindsay firm which agreed to build the plant but the French company La Société des Terres Rares, an indication of the determination of the French to establish their own position in the field. While U.S. firms worried regarding cerium supplies, the Indians made clear their intent to process all available supplies of monazite in their own country and showed little interest in shipping monazite to the United States. This posture was in keeping with Nehru's tendency to stand between East and West, but it did not appear to carry over to beryl negotiations. American representatives broached the issue in New Delhi in May and June of 1949, and in October the work was completed in Washington. The counter for a procurement contract for Indian beryllium was a unilateral declaration of intent by the U.S. government regarding assistance in beryl processing and plant construction and provision of basic scientific training in atomic energy. The appearance of a direct *quid pro quo* was avoided, but agreement was reached only with the decision by the Atomic Energy Commission to relax its security strictures regarding the processing of beryllium.[21]

Though the draft agreement on beryllium was initialed 20 October 1949 by Dr. Homi Jehangir Bhaba, chairman of the Indian Atomic Energy Commission, and by Dr. John K. Gustafson, manager of raw materials operations, U.S. Atomic Energy Commission, its implementation was delayed. Its second paragraph stated that India would sell to the U.S. a quarter of all beryl ore mined over a five-year period beginning 1 October 1950. The following paragraph added that should Indian production reach 600 tons annually, then India would sell a minimum of 400 tons to the United States per year. To this proviso Nehru objected, arguing that Indian production would never get that high and that the Americans would be misled into thinking they would receive more beryl than was possible. Ambassador Loy W. Henderson protested that the Indians had initially talked of having 800 tons in stock and an annual production of 1,000 tons, but had gradually reduced their figures to 400 tons in stock and annual production of less than 600 tons.[22]

Pressed, Indian officials admitted they had been overly optimistic in estimating stocks and production and that they intended to ship beryl ore to other friendly powers. They refused to identify the recipient, but France was implied. Henderson warned of the infiltration of Communists in the French Atomic Energy Commission and suggested that confidence could not be placed in it. To his superiors in Washington, he recommended deleting the offending paragraph, though he saw doing so as seriously reducing the technical advantage of the accord for the United States.

Despite American acquiescence to Nehru's request, anticipated shipments of beryl had not yet occurred by the following April. Efforts by U.S. companies to obtain monazite also proved fruitless, though the government continued to work on their behalf. By this time, the AEC had little interest in purchasing thorium, as there was scant market for what was already available in monazite stored in the United States. Other rare earths were a different matter. It was hoped that the French, while returning to the Indian government the thorium produced by the processing plant being planned for Travancore, would sell cerium and other byproducts to U.S. firms. New discoveries in the United States of monazite sands low in thoria content but rich in other elements were now expected also to ease the problem.

What was viewed as a lack of cooperation by India was a source of some irritation to a group of American congressmen busy reviewing in 1951 proposals for food aid to India. The Atomic Energy Commission also was prompt to remind Acheson, now secretary of state, of the utility of keeping in mind India's recent claim of uranium discoveries and the beryl and thorium problems during the wheat talks.[23] To inquiries from these sources, the department of state replied that the course of negotiations regarding rare earths

> is relevant to United States assistance to India in its present food crisis only to the extent that the embargo on the export of these atomic energy materials reduces India's ability to pay for food grain imports. In that connection, you will be interested to know that if we received as much monazite sands and beryl as we desired from India, the dollar exchange earned annually by India would not exceed two million dollars. This is hardly a significant figure when weighted against the cost of two million tons of grain.
>
> We should not, then, link our need for these strategic raw materials with India's need for food to save the lives of its people.[24]

The Americans and British had shown patience in their rare ores negotiations with the government of India. Acknowledging the stand-offishness of Nehru, they avoided pressing too hard for fear of alienating the newly independent nation. Yet at the same time they would not themselves be pressured into forcing or subsidizing industrial firms in building the processing plant desired in Travancore. The willing intervention of the French again revealed their ambitions in this area, ambitions which symbolized the increasing difficulties that the Combined Policy Committee nations could expect to meet as they attempted to restrict atomic energy development in the free world to their own aegis. The difference between the British and Americans over the utility of attempting preemptive control of monazite deposits was a long-standing one. In the years since the end of the war, the British position had gained increased validity as the importance of thorium remained potential rather than real and the size of monazite deposits became better understood. The Americans were not relinquishing their position in principle, but they were beginning to relinquish it in fact.

Indonesia, Sweden, and Norway

The sensitivity of a newly independent nation and its desire to fall completely in neither the Eastern or Western camp affected also another of the CDT's monazite supply arrangements. Relations with the Netherlands on this matter went well; in 1948 the agreement of 1945 was renewed for another three-year term, with provision made for price adjustments should any purchases actually be made. The Combined Development Agency requested permission to explore monazite reserves on three islands in the Netherlands East Indies, to which the Dutch agreed. At the same time, the Dutch voluntarily produced data which seemed both to meet the Agency's needs and to suggest that the deposits were not of commercial interest.[25]

The creation of the Republic of Indonesia in December 1949 led the Dutch government to notify the United States and United Kingdom that it could no longer fulfill its obligations under the monazite agreement. As far as the territories now under the sovereignty of the Indonesian Republic were concerned, the terms of the agreement had lapsed. Dutch resentment of American criticism of their colonial position in the Far East was substantial, but it was obvious that the small European nation could not produce monazite by itself and that the accord had to go by the boards. Technically this last was not exactly the case, for the December 1948 Netherlands-Indonesian Agreement on Transitional Measures provided that Indonesia assume the rights and obligations of former Dutch treaties where applicable to Indonesian jurisdiction. As the decade of the 1950s began, the American ambassador to Indonesia was instructed to approach the new government at the earliest appropriate time. Given the turmoil in the island nation and the leftist orientation of significant political groups, the wait for resolution of the issue promised to be lengthy.[26]

If the monazite agreements which had looked so favorable at the close of the world war did not work out quite as well as hoped by the Combined Development Trust members in the subsequent half decade, it was of no great matter. More important was the successful procurement of uranium from Belgium and elsewhere and the prevention of that material from falling into the hands of the Soviets. The three Trust nations did indeed believe their efforts had hindered Russian progress. General A.G.L. McNaughton of Canada conveyed reports of miners hand-picking dumps in Czechoslovakia, a sure indication, he thought, of the lack of any real supply in the U.S.S.R. In October 1947 he estimated that it would take the Russians five to ten years to make enough material to produce a single bomb.[27] The estimate of the U.S. Central Intelligence Agency (CIA) that December was not quite as rosy, but still predicted that "it is doubtful that the Russians can produce a bomb before 1953 and almost certain they cannot produce one before 1951."[28] The Russians were suffering shortages of equipment and apparatus. The CIA thought that the Russians would have enough uranium to produce eight to fifteen atomic bombs within the next three to five years; after that, their ore production rate would limit them to one or two per year.

The importance of keeping key equipment and rare materials from the Soviets and of enhancing CDA ore supplies to assure a great preponderance of Western weapons when the Soviets finally did produce an atom bomb took high precedence in the thoughts of American officials. Senator Brien McMahon was sufficiently concerned to propose an amendment to the Economic Cooperation Act funding the European Recovery Program which would bar any participating country from exporting commodities in a manner inconsistent with the national security of the United States as determined by the secretaries of state, defense, and the chairman of the AEC. He was persuaded to abandon this effort by state department officials, who saw it as meaningless in terms of countries such as Belgium with whom ore agreements had already been signed and likely to offend Switzerland and Sweden, two countries which would receive little under the ERP but which were capable of shipping ore and especially atomic energy equipment. Inclusion of such a clause might make them reject even their minimal participation in the ERP and remove themselves a bit farther from American influence.[29]

The posture of the Scandinavian countries and especially Sweden concerned the Americans. In September 1948 the National Security Council called for efforts to strengthen the tendency of Norway and Denmark to align with the West. The Council members also wished

> to make perfectly clear to Sweden our dissatisfaction with its apparent failure to discriminate in its own mind and in future planning between the West and the Soviet Union; to influence Sweden to abandon this attitude of subjective neutrality and look toward eventual alignment with other Western Powers. . . .[30]

In terms of military aid, the Council thought Sweden's requirements should be considered only after those of other countries which had indicated intention of cooperating with the United States or the Western European Union nations (Benelux, France, and Great Britain).

The European affairs desk of the state department braked implementation of the Council's recommendations as far as Sweden was concerned. The report had been written without taking into account the uranium question. The AEC was on the verge of asking Sweden for as much uranium as possible. Sweden's pilot plant for production of uranium from kölm was successful, and the department was now sitting on Swedish requests for equipment which would expand production. Export of this equipment and of defense material should be held up while the AEC examined the issue.[31]

H. Freeman "Doc" Matthews, the U.S. ambassador to Sweden, was asked whether American efforts to purchase uranium would budge Sweden from its neutral position. Circumstances had changed since 1945. Prospects for international control of atomic energy were now minimal. And while the United States could not individually give defensive assurances to the Swedes, their membership in some sort of Western collective defense organization might serve that purpose. Sweden now had the means to process shale for uranium, and she had important equipment for export which on two occasions already she had withheld at U.S. request. Would a uranium purchase offer bring cooperation or a hardening of Sweden's neutrality? Should it be risked? Would a rejection give ammunition to the U.S. claim that Sweden's fence sitting, rather than serving American interests, as the Swedes asserted, was actually harming the West?[32]

Ambassador Matthews and his staff thought that a *sine qua non* for any procurement of Sweden's uranium would be the abandonment of her neutral policy. This was not likely to occur unless the United States was prepared to take "very drastic action" to modifiy Swedish thinking and policy.[33] The ambassador said the Swedes saw the Americans as in no position to give *prompt* military aid if Sweden were attacked. The matter of Sweden's currently gaining U.S. equipment and, if attacked, additional aid was in any case not dependent on whether Sweden were neutral. Matthews recommended therefore that no application for uranium be made; the Americans and British should tell the Swedes, "sadly *not* nastily," that all equipment had to be saved for friends and potential allies.[34]

The atomic energy commissioners soon decided that they held more interest in obtaining information on the Swedish atomic energy program than in procuring Swedish uranium. The safest place for the latter was in the ground. Yet the magnitude of Swedish equipment requests and plans for a full-scale extraction plant to produce twenty to thirty-five tons of uranium per year stimulated questions as to how the Swedes planned to dispose of their stocks. The Soviets were currently experimenting with Estonian shales. Was Swedish technology safe?

Matthews was instructed to work with the British to obtain answers in conformity with Undén's promise of 1945 to keep the powers informed.[35]

Even as the effort to obtain Swedish uranium was set aside for lack of sufficient leverage to move the Swedes from their neutrality, another Scandinavian issue arose which illustrated again the divergent views of the Americans and the British, as well as the burgeoning nature of the atomic energy issue. In August 1948 the Norwegians, endeavoring to construct their own heavy-water atomic pile, asked the British to process some of their pitchblende deposits to produce 10 to 20 tons of uranium oxide for their pile. Having previously turned down a Norwegian request for 5 tons of metallic uranium on the pretext of a raw-material shortage, the British were inclined to reply in the affirmative. Were Britain not to help, the Norwegians declared they would build their own refining plant, which of course would be vulnerable to invasion. The British, moreover, thought it "desirable that the Norwegians and other European governments should look to Britain rather than to France for help and guidance in atomic energy developments."[36]

American officials feared that meeting the Norwegians' request would enable them to ascertain the degree of purity of oxide required for use in a pile, a sensitive piece of information. Then, too, the Russians might occupy Norway and seize the purified oxide. Above all, the Americans thought a reponse should await an overall review of U.S. atomic energy policy. The American joint chiefs of staff additionally cited the demands that would be placed on Norway's limited budget and on world uranium supplies if the Norwegians developed a processing plant. They also disliked the idea of creating a precedent in aiding the creation of any new processing facilities and believed any delay imposed on the Norwegians was worth the risk that the latter might turn to France.[37]

The British remained convinced that the Norwegians should not be rebuffed or forced to wait out an interminably long policy review in the United States. After several attempts to persuade the Americans to withdraw their opposition, London determined to encourage the Norwegians. They would be told Britain would help them with the oxide, but that it would be some while before processing time would be available. Gordon Arneson, the state department specialist on atomic matters, and others muttered; but Under Secretary of State Robert Lovett did admit that any reply which could forestall driving the Norwegians into the arms of the French without committing the United Kingdom actually to take on the work was desirable. In the end, the British avoided any specific agreement to refine Norwegian materials, primarily on the insistence of the United States.[38]

The delay in gaining British assistance may have spurred the Norwegians to turn to Sweden for assistance. The Swedes, for their part, proceeded with their own atomic energy plans and made discreet inquiries about informational assistance. Given the snarl in American-British-Canadian information exchange, Matthews was told, any formulation of policy regarding information sharing with non-CPC countries had to wait. By the end of 1950 the Swedes and Norwegians were discussing an exchange of Swedish uranium for Norwegian heavy water. The Swedes went out of their way to determine the reaction of the United States to this proposed swap. After consideration, the Americans and British declared the matter solely one for internal Swedish determination. The exchange, however, required alteration of Swedish export laws, and the Americans nervously urged the Swedes on several occasions not to allow any loopholes by which the Soviet Union could gain access to Swedish uranium. The cautions of the American joint chiefs notwithstanding, it was evident that a thumb could not be kept in the dike of progress forever. Opposition and foot-dragging would have little positive effect but could cause irritation and turn third countries in a direction neither the United Kingdom nor the United States wished to see them move—toward France. This danger did not prevent the Americans from denying a 1950 request by the Norwegians for purified uranium and reflector graphite, a request which the Norwegians recognized might be denied in view of new security concerns stimulated by the Fuchs betrayal.[39]

Improving Relations with France

Why the dislike of French involvement? Security was the prime reason. A second was the Trust nations' belief that the French would not be likely to harmonize their atomic energy policies with those of the United States, United Kingdom, and Canada. Of course, too, expansion of the French program and French assistance to programs of other countries would mean greater competition for scarce resources.

Awareness and concern about French atomic activities existed during the early war years, but these were allayed by belief that French scientists would resist German efforts to obtain their knowledge and material. This was the case, as Joliot-Curie sent his disciples Hans von Halban and Lew Kowarski to London with a supply of his heavy water and forwarded his uranium oxide to Morocco for hiding.[40] The situation changed when France was liberated. Joliot, a known member of the Communist party, held a dominant position in French nuclear science circles; he could be expected to push for inclusion in ongoing allied research. The British and Americans alike saw him as a severe security risk and an unfortunate influence in Belgium as well as in France.

At a January 1945 meeting of the CPC, Sir John Anderson warned of the need to give some statement to Joliot to postpone the issue and "to protect against political explosion by the French with or without collaboration with the Russians, with possible danger to security."[41] Neither the Americans nor the British wished questions about their project to be raised with France either through Joliot or Charles de Gaulle, leader of the Free French forces. An innocuous statement was agreed upon, yet it was clear that the imbroglio of the preceding months regarding French scientists working in the British project, their claims to certain patent rights, and their desire to return to France to visit with Joliot had strengthened American resolve to have as little contact with the French on nuclear issues as possible.

The Americans' attitude on atomic relations remained standoffish for some months. They disliked Joliot's importuning of Belgian scientists and his suggestions for a European nuclear science pool to which Belgium would contribute uranium. When in 1947 the French approached the British about a patent pool, the latter suggested that the time was not right. They did want to make some broad statement about establishing closer relations in the nuclear field. To this, Edmund Gullion in the state department did not object, but he quickly cited the McMahon Act as preventing any extension of the basis of cooperation with the French.[42]

Yet the French could not be ignored. When Secretary of State Acheson, Secretary of Defense Forrestal, and Sumner Pike, acting AEC chairman in Lilienthal's absence, submitted a special report on atomic energy policy to President Truman in March 1949, the role of other states required mention. Some twenty by then had atomic energy legislation, and over a dozen had atomic energy commissions. Lack of understandings with these countries resulted in confusion and waste.

There is discernable a tendency, potentially dangerous, for the small countries of Europe to band together and exchange with each other their meager knowledge. It may well be, for example, that France will become a center of information, research experience, and development for the "have not" countries.[43]

Joliot-Curie was in conversation with both Norwegian and Swedish scientists, and while the Swedes had been warned of supplying Joliot information which it was assumed would immediately flow to Moscow, the Americans remained anxious. As Swedish foreign office Secretary General Dag Hammerskjöld commented, scientists could be quite naive. The French were asked to clean house, but it was feared that if Joliot were ousted from his post on the French AEC he would take his expertise bodily to the U.S.S.R. Arneson saw a small silver lining in Joliot's continued presence in France, for it offered a ready excuse to deny information sharing. Were he to leave, the French nuclear science community would still be thoroughly "tainted with Sovietism" through the many individuals Joliot had brought into the program, yet a negative response regarding information would then be more difficult.[44]

A key problem in this issue of sharing with non-CPC countries was that because no one knew the extent of the Russians' nuclear sophistication, it was necessary to curtail a wide range of information. Some of the guessing was dispelled with the Russian atomic explosion that summer. At the end of 1949, François de Rose, a member of the French delegation to the United Nations about to return to the Quai d'Orsay as an atomic energy specialist, called on Arneson. The wide-ranging discussion set the tone for better future relations. De Rose planned to to give more centralized focus to French atomic policy. He saw French science divided into hostile and uncoordinated groups, complicated by Joliot's "stranglehold" on atomic energy and his Communist affiliation.[45] The Frenchman hoped to direct the attention of his superiors to the problems and slowly move to coordinate French scientific activity in collaboration with the West. It was to the Americans' best interests for Western Europe to play a role, he argued; he would hope for scientific and personnel exchange with the United States in areas of basic research. Arneson noted that the Soviet detonation had provoked a review of information classification. Much would now be down-graded and could be shared.

No promises had been made, and the Joliot problem remained.

Nevertheless, the air was cleared, and the Americans now knew they had someone with whom to work in the French foreign office. The need for progress was also emphasized by the discovery that the French AEC had signed a contract with Mozambique for delivery of approximately 16 tons of uranium oxide in 1951—this done after the Americans had declined purchase of the same materials because of their limited amount and high price.[46] There were rumors, too, that France was processing small amounts of uranium procured from Portuguese mines outside the control of the Combined Development Agency. The success of the French in landing contracts to build thorium processing plants in Brazil and India, giving them a near monopoly on thorium and its future possibilities as a fuel in breeder reactors, also gave the French a tool by which to impress other nations. The Americans saw the Parisians as ambitious to form a center for European atomic energy and as trying to draw smaller countries into a joint effort. Joliot's contention that Europe was becoming subservient to the United States in the potent new field of atomic energy was attracting increasing attention. The experienced American diplomat and trouble-shooter Robert Murphy saw Joliot as behind the proposal of Raoul Dautry, general administrator of the French AEC, for the establishment of an all-European Institute of Nuclear Physics.[47]

On 28 April 1950, Premier Georges Bidault of France announced the dismissal of Dr. Joliot-Curie as High Commissioner for Atomic Energy, ostensibly for having stated that a scientist who was truly progressive would certainly not allow his scientific knowledge to be employed in a war against the Soviet Union. Alan Kirk, the former U.S. ambassador in Brussels now stationed in Moscow, noted with pleasure the cries of "anguish and pain" emanating from the Soviet press.[48] The encouraged state department, concerned by Soviet achievements and stimulated by the mandate to expand U.S. atomic bomb production, asked its Paris embassy about approaching the French regarding exploitation of uranium deposits discovered by American mining firms in French Morocco and French Equatorial Africa.

Ambassador David Bruce thought that conditions were not yet appropriate. While many French officials might wish to accommodate the Americans in view of the Atlantic Pact and receipts from the Military Assistance Program funded by the United States, domestic opposition to atomic weapons and warfare meant that the presentation of any such proposal before the National Assembly would result in political suicide. Better to leave the materials in the ground and consider them a potential reserve. Morocco posed a slightly easier situation than French Equatorial Africa, as it did not have territorial status in the French Union. Beryl ore was being quietly shipped to America from Morocco with the knowledge of a few officials in the French foreign ministry and AEC; possibly something might be arranged for uranium.[49]

Over the next eighteen months several developments encouraged the Americans. The French revised their AEC statutes and appointed new commissioners; close personal contacts developed between American representatives and officials at the Quai d'Orsay and elsewhere who were determined to exercise new leadership in atomic energy policy. These all led the Paris embassy specialist on atomic affairs, Robert P. Terrill, to write at length to the department in March 1951. He saw little advantage to the United States in the expanding French atomic program, other than a general advance of scientific knowledge. The disadvantages were numerous. France would compete with the United States for uranium, while the United States

> is excluded from direct and indirect access to the potential resources of the French Union at the same time that America is paying for the atomic weapons supremacy that protects France, is opening up many areas of the French Union, and is contributing indirectly through counterpart funds to the budget of the French Government which, of course, includes the Mineralogical Section of the French AEC.[50]

Security risks were involved, for data from French experiments would "almost certainly" be accessible to the Soviets. In a body politic containing so many Communist sympathizers, it was "illusory" to think that ridding the French AEC of Reds and fellow travelers would do the trick, Terrill argued. French diplomacy and support of U.S. policies might be affected. "A growing atomic energy establishment in France might be conducive to the exercise of independent policies in other fields."[51]

Terrill identified U.S. atomic policy toward France as having four cardinal points: (1) maintenance of French support in the United Nations for U.S. proposals for international control of atomic energy; (2) embargo by France of shipment of strategic equipment and materials to the U.S.S.R.; (3) no encouragement of French atomic energy projects joined with informal efforts to turn French research activities from nuclear physics; and (4) discouragement of French activites in the international atomic field and especially technical relations on atomic matters with non-Western European nations. The United States had achieved moderate success in its first two objectives, less in the third, as Marshall Plan aid inevitably strengthened the French industrial base, and very little in the fourth area, where successful instances of limitation of French activities had required direct U.S. influence, not on Paris, but on the third countries involved.

How could the United States dissuade France from "an overly ambitious program?"[52] Maintenance of the *status quo* would leave two objectives unmet. Emphasis on military needs within the Atlantic Alliance and the concomitant suggestion that France not waste her resources nor risk security breaches by attempting a large atomic program might have some impact. But any restrictive obligations would soon become known and arouse outcry in a country where "resistance to American 'pressure' is axiomatic for all parties."[53] No cabinet would dare enter such a deal.

Terrill therefore saw opportunity for the United States to move toward a policy of limited cooperation. Greater support for French research could be granted via more liberal interpretation of the U.S. Atomic Energy Act if France reoriented her policy. The United States could show a more benevolent attitude toward the proposed European Nuclear Physics Center, and the Americans could offer to buy French atomic energy source materials if the French would amend their laws to allow private prospecting.

This approach seemed feasible in Washington, and several discussions were held concerning the inducements the United States might offer in return for opportunity to purchase Moroccan uranium. Meanwhile the French chambers passed legislation encouraging private prospecting and production of uranium. Administrators simultaneously planned to remove Communist employees from the French AEC. The proposed European Nuclear Research Laboratory concept passed into an active stage with support of twelve Western European countries. The prospect of German research activity was coming closer to reality, stirring French apprehensions which they communicated to the Americans. American support of the French position in Morocco in the harsh United Nations debates won important French sympathy and lessened tensions. The French government expressed approval in principle of U.S. access to Moroccan uranium. Thus it was that in the closing days of 1951 Terrill judged the time ripe for specific dealings regarding the Moroccan ore.[54]

The American posture and attitude toward the French atomic energy program had come a long way since 1945. That journey was facilitated, as was so much of United States policy, by the pragmatic need for uranium. Without that need, it is likely that the United States would have maintained its critical aloofness. A new spirit of flexibility appeared, fostered by the expanding number of countries involved in atomic research. Their friendship was desired, and the Maclean and Fuchs betrayals and the Soviet detonation in 1949 had exploded much of the argument of cold war hardliners that every scrap of information should be kept secret. Above all, Franco-American atomic energy relations improved due to the ability of the French to heed American concerns while at the same time pursuing their own objectives. Solid diplomatic work in the lower echelons on both sides, joined with some courageous action by Bidault, made this possible.

The Iberian Peninsula

The American effort to gain access to Moroccan ores was conducted with the approval of the British, who had never taken so harsh an attitude toward the French AEC as had the Americans. Morocco was not the only area to which the Combined Development Agency turned its gaze. Over the years its purview had expanded substantially; its activities in all areas would be both exhaustive and tedious to track. Most searches identified supplies that were not of major interest. One area which drew increasing attention, however, was the Iberian peninsula.

In the middle of 1947 Trust members received reports that the properties held by the Trust in Portugal might have greater uranium reserves than earlier anticipated. Perhaps the previous go-slow policy in the mines' exploitation should be revised. Export licenses would be needed. The British approached the prime minister and quasi-dictator of Portugal, Dr. Antonio de Oliveira Salazar. The Portuguese were offered opportunity to reserve a reasonable amount of ore for themselves and to levy reasonable export taxes on the remainder shipped over a ten-year period. The talks dragged, but by 1949 the negotiators had achieved a contractual agreement permitting export of uranium from mines held by the Combined Development Agency. The agreement covered just seven years; on the other hand, the Portuguese government promised to do all it could to prevent uranium from its territories reaching " 'persons inimical to' this understanding, i.e., persons within the Soviet bloc."[55]

Problems did arise. High import duties on structural steel for the plant Britain began at the Portuguese Urgeirica mine slowed construction until they were finally lowered after several protests.[56] Production grew only slowly; ores from non-CDA mines often moved to other European countries such as France. Salazar seemed only vaguely aware of the American interest in the ores.

Thomas E. Murray, a member of the United States AEC, called on the prime minister in October 1951 to stress the urgency of the situation. Salazar immediately made known his wish to hear from the Americans directly rather than via the British. He believed that Portugal was not receiving a proper price for her uranium, but rather only the cost of getting it out of the ground. Unaware of the workings of the CDA, he may have hoped for a higher bid from the Americans than from the English. Salazar intimated further that price was the reason output was limited. He inquired, too, about industrial and medical uses of uranium. Murray avoided any promises but did warn the prime minister of the importance of uranium for the safety of the world. Portugal's ore would be of little or no value in 1960 if events turned out badly. If he wished to protect world peace, now was the time to act.[57]

Murray also called on Generalissimo Francisco Franco, premier and chief of state in Spain. Again he emphasized the uniqueness of uranium compared with other materials and the significance of time. It took some while to turn ore into bombs, and 1960 might be too late. Preliminary negotiations had been begun with the Spanish government through Westinghouse Electric International. Murray asked for an expedited survey of Spanish resources plus an agreement that Spain would export all her uranium to the United States, most likely at a price of cost-plus-reasonable-profit.[58]

Franco was receptive but had reservations about a declaration of intent to sell to the United States. He doubted that Spanish deposits of uranium were significant. Having already met with the Spanish foreign minister, the following day Murray and a Westinghouse representative visited with the Spanish minister of industry, who urged that more geologists than the two or three suggested by Westinghouse be sent to the scene. Murray and the chief of an American military survey team then in Spain also met with interested Spanish military officers.

Two months later the Americans returned to the issue of a declaration of intent on the part of Spain. They had and would again make vague statements about technical assistance and personnel; yet they emphasized that these were contingent upon the United States receiving uranium. Franco was still unwilling to sign a declaration of intent to export to the United States all uranium in excess of Spanish needs, although his minister of industry had already prepared a draft to this effect. The dictator insisted that his unwillingness to proceed was due to his country's lack of knowledge of uranium development. More study was needed; meanwhile he would welcome American engineers and assure that no Spanish uranium would be sold to a Communist power. He intimated that he had been approached by several states, including Belgium, about Spain's excess production. The American ambassador suspected, however, that Franco simply wished to tie uranium matters to the program of military aid soon to be discussed with a United States defense mission.[59] Franco had forced Hitler to bargain hard for Spanish mine products in time of war. Why should he act otherwise in time of peace?

Negotiations with South Africa

The value of residue ores produced in South African gold-mining ventures was ascertained not quite by accident. In his journeys Professor George Bain had determined that uranium did exist in the gold mines of the Rand. Subsequent investigations by others suggested this presence was of insufficient richness to be of use. This finding conflicted with Bain's own calculations. When he demonstrated that a souvenir sample of gold-bearing rock emitted beta rays such as to promise valuable deposits, a new inquiry was launched upon the insistence of Groves and with the backing of Sir Charles Hambro and Dr. James Chadwick on the British side. The uranium, it was discovered, was located in a thin stratum within the larger deposits of gold-bearing rock. Normal extraction methods left this special ore mixed with other slag in low concentrations of about three-tenths of one percent.

Early in the uranium search, recovery would not have been possible. But in the postwar years Dr. A. M. Gaudin of the Massachusetts Institute of Technology had been working on concentration methods for low-grade Congo ores. He turned his attention to the South African challenge and surprisingly soon developed a small batch process for dealing with these ores. Further research developed mass-processing techniques, and as knowledge developed about the nature and location of the uranium-bearing stratum, other steps could be taken to assure better concentrations with which to begin. South African uranium exports were to rise rapidly, and by 1959 were valued at $150 million. In addition, previously unprofitable gold mines could economically be reopened when the proceeds from uranium sale were also taken into account.[60]

The cooperation and help of both the South Africans and the British in uranium procurement was a frequent concern to Lilienthal and others during the discussions about information sharing with the British. Indeed, it was evident to the Americans that the British were not at all eager to let South Africa be considered a territory to be dealt with by the Combined Development Trust. Rather, they wished to keep control of South African ores themselves and use that control as a bargaining chip for information sharing.[61] As part of the ill-fated November 1945 agreements and then of the *modus vivendi* of January 1948 it was agreed that Commonwealth ores would be administered by the Combined Policy Committee and contracted for by the Combined Development Trust.

Throughout the stages of exploration and initial development, the prime minister of South Africa, Field Marshal Jan Christian Smuts, proved friendly and supportive. He recognized the importance of moving quickly and was willing to cooperate with the United States and the United Kingdom. South Africa would retain control of her uranium and, he thought, supply it only to the British and Americans.[62]

This smooth sailing was threatened by the June 1948 electoral victory of the Nationalist party led by Dr. Daniel F. Malan, which thrust Smuts and his Union party from power. Malan's vehement nationalism and separatism threatened cooperation between Capetown and London. Yet the nationalists actually trailed the unionists by over ten percent of the popular vote and had a majority of only a few seats in the South African parliament. They too needed the economic benefits uranium export would bring, and Malan was an outspoken critic of Communism. In this regard the Trust had word that the new government "would be anxious to place its material in the hands of those who could make best use of it in the fight against communism."[63]

The talks regarding price, quantity, and a commercial contract begun prior to the elections continued. The anti-British attitude of

Malan made it politic for the Americans to take the lead in the negotiations. Their members of the CPC contemplated a new pricing arrangement because of lack of knowledge of the costs of the new extraction process and because of the wide range in concentration of ores. Under Secretary of State Robert Lovett urged that political inducements should not be raised unless the South Africans did so.

> In the last analysis, however, the price of uranium was dependent on political considerations. The South Africans, like the Belgians, were willing to sell uranium because they felt it is in their interest politically and strategically.[64]

Lovett estimated that since the South Africans had asked for bids from the United States and United Kingdom and made it evident that their chief interest was assurance of sales of uranium over the long term they would confine their talks primarily to financial considerations. In the long run the United States would probably have to acquiesce to the existence of a nuclear pile in South Africa, but initial assistance should be limited and informational security made certain. Carroll Wilson, general manager of the AEC, warned that "In all our future foreign relations with The Union of South Africa we would have to bear in mind the importance of South African uranium."[65]

The exploratory talks of June 1948 led to a contract for 10,000 tons of contained U_3O_8 in a high-grade concentrate. The price would be fixed at cost-plus-profit and royalty. The seller would have option of delivering up to 1,500 tons of the total at a rate of 150 tons per year at a price not exceeding $25 per pound of oxide. The balance of the contract would be filled at a minimal rate of 400 tons per year sold at a price not exceeding nine dollars per pound. Payment would be half in dollars and half in sterling, in accord with established CDA policy. To the South Africans so short of foreign exchange that limits on imports would soon become necessary, this prospect was attractive. To be sure, more research and pilot plant experience were still needed; yet there was hope that uranium production might begin in 1952.[66]

Though all appeared well, Gordon Arneson remained concerned that a falling out of the United States and the United Kingdom over information sharing could sour the negotiations, for as time passed the Malan government seemed to be willing to move slowly back toward the Commonwealth family.[67] This too was a concern of Acheson, Forrestal, and Pike which they later aired in their March 1949 special report to Truman. Any termination of the Combined Development Agency relationship would jeopardize potential receipts from South Africa. Something had to be worked out on more flexible information sharing also because countries such as Belgium, South Africa, and Sweden would not long be content without their own research programs.[68]

Final formal agreement with South Africa had been postponed until more knowledge of costs and production expectations had been gained. Though plans were initially made for a joint U.K. and U.S. mission to South Africa in the spring of 1949 to seal the bargain, negotiations were delayed for various reasons. As months passed, the possibility of a tri-partite government pact to back up any commercial agreement was considered. While the state department was pleased that it was being successful in keeping the uranium issue separate from other matters, Americans began to monitor more closely other issues: South African disappointment regarding the nature of financial assistance from the Export-Import Bank, South African desire for an increase in the price of gold and for U.S. military aid, and touchy problems such as treatment of the races in South Africa which were again coming before the United Nations.[69]

In November 1949 the South African Atomic Energy Board and representatives of the Combined Development Agency reached an agreed basis for subsequent negotiations. Yet a formal signing had still not occurred by March 1950. Expectations that South African ores would be the key resource after the exhaustion of Congo deposits, together with word of the arrest of Klaus Fuchs in February 1950, pushed U.S. Secretary of Defense Louis Johnson to the point where he could brook no further delays. Moreover, he had long questioned the policy of working and sharing information and materials with the British. On 13 March he wrote Acheson that recent disclosures and other uncertainties were compelling reasons to make sure the results reached in talks with the South Africans were "precisely tailored" to U.S. needs.[70] He recommended a straight business transaction involving the United States and South Africa only. Ore should be stockpiled in the United States. No third party had an inherent right to participate in allocations of that ore, and prices granted to third countries should take into account the American investments which made any proffer of ore possible at all. This last comment referred to capital invested in developing a means of processing South African ores and building a pilot production plant.

Acheson let the memo cool, then rejected it, citing the response given by Sumner Pike, acting chairman of the AEC following Lilienthal's departure. To accept Johnson's advice would require a reversal of current negotiations with the South Africans and would prejudice the existing good prospects for agreement. It would also reverse long-standing policy on materials acquisition and would conflict with American obligations under the *modus vivendi*.[71]

The American members of the CPC concurred and also agreed on a manner of meeting a new issue raised by the South Africans, namely a request for a "special position" in the field of atomic energy. The approach had been made to the British; their recommendation was that some provision similar to that granted Belgium by Article 9(a) of the Tripartite Agreement of 1944 might be allowed, though it should be independent of any contractual negotiations. In due time the South African ambassador also asked the department of state that his country be included in the inner circle of Western atomic powers in view of its future contributions of uranium. There was no need for concern about security, he said; his government would be careful. Besides, it had no interest in information relating to the manufacture of weapons.[72]

The South Africans chose not to press the issue of their status in the atomic energy club. Indeed, they purposely held it in abeyance until a "Memorandum of Heads of Agreement" dealing with contractual arrangements between the CDA and the South African Atomic Energy Board was at long last signed in Johannesburg on 12 November 1950. Acheson expressed gratification over the successful conclusion of the contract negotiations and assured South African Minister of Interior Dr. T. E. Donges that "something could be worked out" on his request, not for full membership but, as Donges put it, "associate membership in the club" of atomic powers.[73]

The Americans and British believed that while South Africa could not be admitted to the CPC, which had been formed under special circumstances during the war, she could be entitled to the same special position held by Belgium under the 1944 agreement.[74] Neither wished to rush the matter. Any position South Africa would be given could not exceed that granted to Belgium. That country was still the Combined Development Agency's chief current supplier of ore, regardless of South Africa's long-range potential. Were the Belgians to suspect that special privileges were going elsewhere and being denied to them, serious repercussions would be sure to follow. The desire of the South Africans for inside information demonstrated once again that the CDA's efforts to obtain ore were increasing the dimensions of proliferation of nuclear capabilities by diplomatic bargaining. Scientific advances in several countries fostered this growth, in any case, yet the issue would not have grown so rapidly had the British and Americans initially possessed their own sufficient deposits of uranium or had they discarded the notion of preemption.

Now the CPC and CDA were engaged in some of their most difficult negotiations of the postwar period with the Belgians. Failure in these would make all the other efforts meaningless. The readiness of the World War II defenders of democracy to dance with the likes of Vargas, Salazar, Franco, and the South Africans would count for nought except embarrassment if the continued collaboration of the earliest, most faithful, and highly democratic ally could not be assured.

8 · Raising the Compensation: The Belgian Export Tax

Wisely or not, the West in the years following the German and Japanese surrenders rested its defense and security primarily upon the atomic bomb. As sole Western possessor of the bomb during this period, the Americans found themselves in a unique position in terms of responsibility to produce the bomb, guarding the safety of their friends and allies, and in their dependence upon the cooperation of one of the smallest of the powers. That cooperation, moreover, rested more upon the will and understanding of two men, Paul-Henri Spaak and Edgar Sengier, than upon the diverse political establishment and popular will of Belgium. This last was emotional and far from stable, wracked and divided as it was by economic problems, accusations regarding posture and performance during the unexpected rout of May 1940 and subsequent German occupation, bitter linguistic controversy that vastly complicated the political divisions, and the consuming debate over the proposed return of the self-exiled King Leopold III.

All these issues and more demanded sure and skilled footwork by Belgian politicians if they wished to remain in office. Belgians had never lacked pride in their country, which had been established in such difficult circumstances from 1830 to 1839 and which had persevered against high odds. The tutelage of the great powers had been alternately and even simultaneously appreciated and resented over the years. Prestige won by the sacrifices of 1914-1918 served as encouragement in the dark days of 1940-1944. However, unfortunate and rash accusations of treachery and cowardice thrown the Belgians' way following their shockingly rapid defeat of 1940 shook their national pride. Desire of recognition for important service performed in the delivery of such crucial material as uranium would soon join with concern that the significance of this service might not be appropriately rewarded. Fear rose within the country that it might again be short-changed as most Belgians believed it had been at the Paris Peace Conference of 1919.

The pressing nature of so many issues other than uranium and the very character of Paul-Henri Spaak enabled the Combined Policy Committee countries to prevail upon the Belgian premier to give the September 1944 accord as low a profile as politically possible in Belgium. Spaak was committed to broad international cooperation and to the defense of the West against the Soviets, regarding whose maneuverings he was one of the first European spokesmen to warn. Yet even Spaak was subject to political pressures which led him to importune the United States and England to permit some public statements which might allay rumors that Belgium was being taken advantage of, or that she had sold her most valuable Congo inheritance for a mess of pottage. In fact, it had been sold at what at the time was considered a reasonable price, a price that involved thousands of dollars and pounds sterling. More important to the far-seeing and future-oriented Spaak, the sale of uranium meant that Belgium would be in a special position in terms of promised technology that would put the country at the forefront in the continental development of what was imagined to be an energy revolution restructuring the entire industrial future.

Despite Spaak's pleadings, the British and Americans let him say little; revelation of the duration and terms of the accord might give the Russians information better kept secret or might play into the hands of the domestic Communist opposition in Belgium. As a man more concerned with substance than statements, Spaak was not truly convinced that Belgium was, in fact, receiving the sort of technological information she had been promised. The reluctance of the British and especially the Americans to provide this information was tied to the close linkage of weapon-related knowledge and that for industrial energy. Equally significant were the American fears for secrecy, which in hindsight seem both foolishly hysterical and yet justified in view of the betrayals of Maclean, Fuchs, and other spies. The tight-lipped behavior stimulated by these fears was accentuated by American pride of discovery and of effort which vastly complicated the concept of the information-sharing ideal Churchill and Roosevelt had so happily embraced at Quebec. If the British, Americans, and Canadians could not work out a mutually acceptable system of information sharing, how were they to fulfill their obligations to Belgium?

Belgian Pressure for Information

President Truman's specially appointed policy subcommittee of the National Security Council recognized the need to take some steps to improve relations on atomic matters with friendly non-CPC countries. Early in 1949, Acheson, Forrestal, and Sumner Pike, acting chairman of the AEC, endorsed their staff's suggestion that foreign scientists be invited to work in American universities and unclassified laboratories. Assistance should also be given to the development of acceleration projects and other nuclear science activities unrelated to making fissionable materials or weapons. The subcommittee opposed the development of atomic energy outside the three CPC nations but agreed that special cases raised by countries providing raw materials should be individually decided on their merit. Assistance rendered should be kept at a minimum consistent with general foreign policy and the acquisition of materials. Should assistance be given at such time as to reap good will without making any significant contribution to the production of atomic energy in another country, or should the information be withheld on principle, consonant with a clearly enunciated policy of only minimum assistance? The three believed that the first course might give real advantage to the U.S. cause.[1]

In late summer 1948 arrangements were made for Belgian scientists to visit the United States, enroll in the radioisotope school at Oak Ridge and in various universities, and to have greater access to radioisotopes for medical research. These small steps might have been sufficient to satisfy both Belgians and Americans for a number of months had not other developments intervened. The July 1949 Blair House meeting at which Truman and Eisenhower tried to persuade dubious members of congress that they should accept the special advisory committee's recommendations on information sharing received considerable publicity. In Belgium, newspapers took up the question of what was happening to Congolese uranium and published mostly inaccurate figures regarding tonnages and the supposed duration of the agreement with the allies.

At the same time a ministerial crisis resulted in the replacement of Spaak's Socialist party by the Liberal party in coalition with the Christian Social party. On 18 August the new prime minister, Gaston Eyskens, in reply to Communist questioning stated: "No secret treaty exists with any foreign country whatever for the purpose of delivering to it uranium or any other raw material. My declaration is forthright and formal."[2] The Belgian press, aware of Spaak's admission of July 1947 that wartime arrangements placing ore at the disposal of the United States and United Kingdom were still in effect, jumped to the conclusion that the secret agreement had expired as of 31 July 1949 and that a new one would now be negotiated. In fact, the ten-year period following completion of the first deliveries envisioned by the 1944 accord would not end until 6 February 1956.

In Washington, London, and Brussels, those in the know were taken aback. Edgar Sengier and others concluded that Eyskens was making a play on the word "treaty" as distinguished from "agreement." It was inconceivable that he was not aware of the deal, although Sengier did allow that only Spaak knew all the details of the secret agreement. As a matter of fact even the former premier was not acquainted with the contract terms, which Sengier kept to himself. Pierre Wigny was continuing as minister of colonies, but he too knew only the basic outlines.

Sengier estimated that the key to the uranium issue would be whether the new government would have the confidence in him that Spaak did. To bolster his position vis à vis that government, the businessman requested a letter from the Combined Development Agency assuring that the price paid for Belgian uranium was *"as good as"* the price paid other suppliers.[3] Such a letter was quickly prepared, for the Americans viewed the essential factor to be profit, and this Sengier was making.[4]

The Americans did suspect that in October Sengier would request a price increase in order to gain a more defensible position with his government. He had said he thought prices paid so far were satisfactory, and this point would soon have to be made clear to the new government. The politicians also needed to be assured that Belgium was not involved in the alleged "row" between the United States and the United Kingdom which stimulated the Blair House meeting. Finally, Eysken's cabinet must be convinced that Belgium's interests were fully protected, in light of her diminishing coal supplies and the potential for uranium-derived energy. Gordon Arneson again thought that Spaak's 1947 statement that the two governments would admit Belgium to participation in commercial exploitation of the ores should suffice on this point.

The assurances proffered in each of these areas might indeed have been sufficient to quiet the concerns of the new cabinet but for two additional factors. The new foreign minister was Paul Van Zeeland, an outspoken supporter of European cooperation and a Catholic party technician who twice had served as prime minister in the mid-1930s. A supporter of the king, he was deeply involved in efforts to arrange an amicable return of the monarch to the country. Out of power for some twelve years, Van Zeeland led American diplomats to believe that he wished to get more out of the uranium agreements in order to enhance his political prestige.[5] A lawyer who had also studied economics at Princeton University, he considered himself equipped to deal with both the issues and the Americans.

The second factor was Truman's announcement on 23 September 1949, while Van Zeeland was visiting the United States, of the explosion of an atomic bomb by the Soviets. When the foreign minister had met with Acheson a week earlier, Van Zeeland suggested that discussion of the uranium issue be postponed. Word of the Soviet success stirred opinion the world over, however, and meant that Van Zeeland would be deluged with questions upon his return to Brussels on 6 October. Belgian Ambassador Baron Silvercruys was therefore instructed to present a *note verbale* to Arneson, accompanied by a request for prompt reply. In it, the Belgians stated that they had fulfilled the provisions of the 1944 agreement but as yet had received no precise indication as to how the two governments intended to fulfill their obligations under Article 9(a) regarding the sharing of information for industrial purposes. Despite promises of determination to aid Belgium, nothing had been done; on the other hand, the American Atomic Energy Commission had as early as 1947 announced a program for participation by American industry in the production and utilization of atomic energy and had established an Industrial Advisory Group. Other information culled from the American press was also cited to suggest that development of industrial applications of atomic energy was underway in the United States but that Belgium was not being informed although her uranium was the *sine qua non* fuel. Van Zeeland also suggested that some joint declaration be made regarding the 1944 agreement, for the discussion in the press demonstrated its existence was no longer secret.[6]

The approach by the new cabinet had been made only to the Americans, with inquiry if the British should be involved. The Americans quickly advised that the latter be contacted. Yet the course taken by Van Zeeland was not entirely coincident upon his presence in the United States. It also reflected how the lead responsibility for uranium negotiations had passed from Britain to the United States in the eyes of most observers and how the responsibility for fulfilling reciprocal obligations now fell primarily on the Americans as well.

Arneson's response, assented to by the British, did not go far, but it was enough to satisfy the Belgian foreign minister for the time being. Contacts with American industrialists were intended, first, to broaden industrial participation in weapons production and, second, to make available new technology which might have application in general industry. Neither of these purposes related directly to Article 9(a) or the development of industrial atomic energy. The Belgians were reminded that it would be years before industrial atomic energy was obtained and that because of the international situation the primary focus of American technological development was still weapons. The United States Atomic Energy Commission would be willing to meet with Belgian technicians to discuss what further could be done, and the AEC would also soon be considering with Sengier the one area of well-developed non-military application, radioisotopes, and how they might be distributed in Belgium. As for publication of the agreement, the United States still clung to the Spaak phraseology of 1947, insisting that any disclosure of the time span of the agreement on the ultimate tonnages of ore involved would negatively affect the security of the three countries.[7]

Van Zeeland desired to make sure that the Americans would accept more meetings of technicians and scientists. He found a point of leverage unanticipated by the Americans when he informally coupled this point with a request to Lewis Strauss of the AEC that, since Belgian uranium was at stake, Belgian representatives be present at any future Anglo-American discussions of uranium allocation. Under Secretary of State James E. Webb quickly told Silvercruys that these talks were in recess and that Belgium would be kept informed of any significant developments. He just as quickly agreed, however, on a manner for designating the Belgian officials who would participate in the projected technical meetings.[8]

Before the Belgian delegation arrived at the beginning of 1950, the Combined Development Agency had successfully completed its contract negotiations with Sengier. He was satisfied with the price, although the Americans wondered if he may have wished an even greater increase so that he would not be subject to criticism from Van Zeeland. He had also received a number of radioisotopes for distribution to medical and scientific programs in Belgium.

Sengier by policy kept price details to himself as much as possible. He went to considerable lengths to impress upon the foreign minister that the 1944 accord consisted of three parts: opening and closing sections dealing with governmental relations and a middle portion on contracts, which should be Sengier's prerogative alone. On 5 January he warned the Americans that:

> the question of price will remain a permanent difficulty. One
> could discuss for years to come what is a "fair price" to be paid

for such rare valuable material. My attitude has always been . . . that we should get a price not lower than price paid to other suppliers for contracts involving more or less comparable top wages. For instance, Canada and possibly South Africa.[9]

While wages for the chief white engineers at Shinkolobwe, Eldorado, and the Rand may have been somewhat similar, surely the wages for the common day laborers were not, and these made up the greatest amount of the Canadian payroll. Thus the price issue could remain a subject of interminable debate, for by Sengier's formula the Union Minière's profit margin would be far greater than that obtained at Eldorado, and the CDA policy continued to be that producers should enjoy similarity of profits rather than similarity of price.

Sengier and his deputy Herman Robiliart had for some time, with the quiet support of Arneson and Lilienthal, tried to hold back those who favored the construction of an atomic reactor in Belgium. An ardent anti-Communist, Sengier did not want any classified information leaked to the Russians. The explosion of the Soviet bomb weakened that argument. Arneson believed that Sengier continued to drag his feet regarding the Belgian reactor, but this may not necessarily have been so. In 1944 Sengier had definitely shifted to a more nationalistic viewpoint once the apogee of crisis had passed, and this may have happened again. On the other hand, Sengier was determined not to allow Van Zeeland to interfere with his contracts, which dictated that all Shinkolobwe production go to the CPC countries. In any case, Sengier did clear his position with Van Zeeland and Minister of Colonies Wigny and then sent Robiliart to participate in the Belgian technical delegation, primarily to protect Union Minière interests.

Failure of the Winter 1950 Discussions

The talks which began in January 1950 did not go well. The preceding month the Belgians had prepared a five-point agenda. First to be considered were methods by which Belgium might benefit from progress in the industrial utilization of atomic energy. Second, means should be determined for associating Belgium actively in research to the extent security and military secrets permitted. Third, Belgium should be associated "in all negotiations having to do with the use and distribution of the ore among the contracting parties of the 1944 agreement."[10] Fourth, there should be an increase in the unit price stipulated in the contract, the surplus being placed in a Belgian public-interest fund. Fifth, agreement should be reached on a declaration of the results of the negotiations.

The agenda was heavy, and its tone and the points raised suggested an aggressive attitude on the part of the Belgians. While the state department was sympathetic on the first point, there was little it could proffer because industrial application was still so distant. Surely it desired no involvement of Belgium in the allocation of the ores between Britain and the United States. It no doubt suspected that Belgium not only would see opportunity to play an influential role as broker but would also demand increasing portions of ore for herself. Silvercruy's comments about strengthening Belgian science in the atomic field were thought to smell dangerously of ambitions to build a reactor, with the concomitant security risks and additional demands on the limited ore supply. Any revelations of the amounts of ore needed by the British and American programs would automatically be a tip-off regarding the extent of weapon production, information the allies scarcely wished leaked to the Soviets by Belgian scientists with leftist leanings.

The state department fully believed that "no political considerations and public relations of [the] Belgian Government (specifically Van Zeeland's political position) outweigh economic aspects. . . ."[11] U.S. Ambassador Robert Murphy in Brussels concurred, but thought that an increase of unit price would "go a long way toward ameliorating political aspects."[12] The creation of a public-interest fund would of course be politically attractive in Belgium. Its inclusion on the agenda, joined with Sengier's warning letter of 5 January that price could be argued *ad infinitum* and that Congo ores should receive the same price as Canadian, suggests that on this point Sengier may have worked out a *modus vivendi* with Van Zeeland and Wigny.

By 10 January 1950 Murphy was a bit concerned about Sengier's posture. A year previously the Belgian executive had been content with his price and profit, but since then he had learned of the extensive efforts made to obtain South African uranium at higher cost. He knew that in a few years he would no longer be playing a principal role. Having for so long taken pride and pleasure in his key part in Western security, Sengier might well now feel hurt pride, and he might believe that perhaps he should look for more profits, if not on behalf of his company, on behalf of his country. Joseph Volpe also found the Belgian more difficult and that Sengier felt betrayed by the British and American failure to provide significant scientific information to his country.[13]

Neither the Americans nor the British liked to be held up on the matter of price, for they were aware of the extensive expenditures they were making which contributed to the security of all of Western Europe. Granted, they made these to protect their own security, yet the French, Belgians, Dutch, and others were equal beneficiaries. Domestic political needs could be understood, however, and so some sort of financial salve might be acceptable as the easiest way to deal with a sensitive situation. As Murphy pointed out, reports of reactor construction in France, Norway, and Sweden made the Belgians, the chief suppliers of uranium, feel they were missing the boat. John K. Gustafson, director of the AEC's raw materials operation, when visiting Brussels at the end of 1949 had suggested that the lower price Belgium received for her uranium could be counted as her contribution to the Mutual Defense Aid Program by which the United States was helping to fund the military preparedness of friendly European states. Logical as this proposal sounded, Murphy noted that it probably would "not adequately meet Van Zeeland's political problems and aspirations."[14]

The negotiations began in Washington on 30 January and continued for three additional sessions through 9 February. Though the talks were presumably technical in nature, the delegations included numbers of policy-making figures. Among the Belgian participants were Ambassador Silvercruys and Fernand van Langenhove, the permanent Belgian representative to the UN, Robiliart, and Professors R. Ledrus and A. de Hemptienne. The key British delegate was Sir Derick Hayes from the Washington embassy. The Americans were led by Assistant Secretary of State for European Affairs George Perkins, Gordon Arneson of the state department, Carroll Wilson from the AEC, and Robert LeBaron of the defense department. Sumner Pike and Henry Smyth of the AEC attended some sessions. Though the Americans entered the talks in the belief that some arrangement could be established on the basis of a reasonable price increase and a reasonable sharing of information, differences about what was reasonable soon disabused them of these views.[15] Their attitude in some ways may have been condescending. They certainly failed to recognize the Belgian determination no longer to be satisfied by palliative formulas and promises.

The Belgians wished to establish their own reactor and desired information which would assist them in this effort. For security reasons neither the Americans nor British were willing to pass on this information.[16] Moreover, the American Atomic Energy Act directly forbade such transmission of sensitive information. If Ambassador Murphy thought the United States should consider giving Belgium help with the reactor, this was not likely to have been the view of LeBaron, one of the chief opponents of information sharing with the British. How could the Americans give information to Belgium that they were denying even to their partners in the Quebec Agreement and in the *modus vivendi?*

At the 9 February session, the American and British representatives presented to the Belgian delegation a "Memorandum of Conversation" outlining the sorts of unclassified assistance they were prepared to extend to Belgium at that time. The Belgians' frosty reaction at the meeting was transformed into a heated memorandum delivered on 14 February. Their patience exhausted, their sensibilities aroused by the thought they were being maltreated by distainful great powers, and no doubt consciously or unconsciously motivated by the traditional small-power view that justice resides with the weaker, the Belgian delegation expressed itself sharply. What was said at the meeting has not yet been made public, but the Valentine's Day memo triggered a negative response within the American delegation to what was considered "tactless" phrasing.[17] They especially resented the impression given that the Belgians doubted the sincerity of American statements regarding the continuing remoteness of any practicable use of atomic energy for commercial purposes. The result of the meetings thus was not progress toward a better and new understanding between the powers but a greater rift.

The Valentine's Day memorandum was soon followed by a strong, though more politic, message from Van Zeeland to Acheson. Suggesting that the negotiators had not given sufficient attention to political aspects of the problem and offering the polite excuse that this was only natural, given the technical nature of the talks, the Belgian foreign minister requested the secretary of state's direct intervention. He assured the American of Belgium's spirit of cooperation and pointed out that the 1944 agreement should have been revised at the war's end and would have been but for the extended and ultimately fruitless hope that an international agreement under UN auspices could be worked out. Five years after the armistice, Belgium, though the world's chief supplier of the key fuel, was one of the Western European countries least advanced in atomic technology. Why? She had relied on receiving key information under Article 9(a) of the Tripartite Agreement of 1944, information which had not been forthcoming. Meanwhile she had rejected an invitation to participate in research on a French atomic pile in return for uranium and had also not considered a Norwegian approach involving exchange of heavy water for uranium. Belgium desired compensation in technical matters for her deliveries of ore.[18]

On the domestic scene, Van Zeeland continued, the Belgian parliament could no longer permit silence on the matter. Interpellations were scheduled. The constititution required the government to inform the chambers of treaties as soon as security permitted; those treaties were not thenceforth supposed to be binding until approved by the chambers.

> These provisions face the Belgian Government with obligations it can no longer evade. The mere publication of the text of the 1944 agreement would place the Belgian Government in a politically untenable position. The Belgian Government can obtain acceptance of the obligations undertaken under this wartime agreement, which has now been continued for almost five years after the end of that war, only on evidence that the continuation of this situation had not been undertaken without compensation.[19]

New Efforts Toward Publication

Van Zeeland saw his request as "reasonable, moderate and of a nature to serve the common interests of the United States and Belgium."[20] Whether Acheson agreed with this assessment is debatable, but he was convinced that the Americans had to show more movement from their previous position. Space for such movement was difficult to find, given the Belgian desire for scientific information and the limitations posed by the McMahon Act. Acheson therefore turned to a matter frequently raised by Spaak but which had not been so prominent in recent talks until mentioned by Van Zeeland—publication of the 1944 accord.

First, Acheson assured the foreign minister of the U.S. appreciation for Belgian cooperation. He also reminded the Belgians that the atomic weapons produced protected all the North Atlantic Treaty Organization countries, including Belgium. Belgium contributed ore, the United States contributed scientific manpower, technology, industrial organization, and expenditures of more than five billion dollars. Security was collective.[21]

Second, the secretary explained that the U.S. program was still primarily focused on weapon production. Development of power piles was so intermingled with classified military work that little could be separately identified and released. Mindful of Belgian contributions, the United States and United Kingdom would do all they could "to place Belgium in a position to take advantage of commercial applications as they develop in the future."[22] Concerned by reports that Van Zeeland was convinced by what he cited as "unimpeachable sources outside of Belgium" that the two powers were farther along in commercial developments than they admitted, Acheson warned Murphy that if the Belgians could not be dissuaded of this view then any meeting of minds would be unlikely.[23]

In his 22 February message, Acheson proposed a statement which he suggested Van Zeeland could use in a forthcoming interpellation. Not the sort of brief avoidance of revelations of significant information characteristic of previously recommended public relations releases, it was a full-fledged communiqué which cited the precise wording of Article 9(a) and (b) of the 1944 agreement and outlined a positive policy for information sharing. In particular, the powers promised to assist in the placement of Belgian students for advanced study in American and British universities, to facilitate Belgian access to declassified material and further visits of Belgian scientists to unclassified work undertaken by the AEC and the British Ministry of Supply, and to assist Belgium in gaining equipment for research in atomic-energy-related fields. They would arrange for Belgian participation in exchange of information on the locations and recovery of radioactive ores and for closer consultation in the distribution of Belgian ores between the United States and the United Kingdom. In addition, Murphy was privately instructed orally to mention the possibility of the declassification of the design for the standard low-power research reactor in which the Belgian technical delegation had expressed interest. He was also to suggest the possibility of U.S. financial aid for Belgium's research once she had clearly defined the program she wished to undertake.[24]

The communiqué, with which the British concurred, seemed a major breakthrough; given the restrictions of the 1946 Atomic Energy Act, it went about as far as could be hoped. That Acheson saw it as a major step is evidenced by his suggestion that Van Zeeland fly to the United States for a formal tripartite signing of the document. It would be better to have an agreed-upon and precise release of the important phraseology than for innuendoes to be cast in the 7 March interpellation; Eyskens' earlier statement had, after all, been quite misleading. An agreed statement to which Van Zeeland was party would offer more long-term security for continued delivery of Belgian ores than a possible whipping up of Belgian parliamentary feeling if Van Zeeland chose to present himself as a David doing battle with a selfish Goliath. Declassification of the reactor information made sense in view of the Soviet advances, and a little acceleration of the process could gain good will. As for the "considerable" financial contribution, it had long been recognized as a probable necessity. Since Van Zeeland was casting the issue in terms of principles and the fulfillment of Article 9, some way had to be found to proffer him the money the Americans thought he really wanted without appearing to buy him off. The ideal route was that of funding the Belgian atomic research program.

If Van Zeeland's immediate concern was the impending interpellation, the proposed communiqué was a bold step and one he had to accept. This he did, pending formal approval by the Belgian cabinet. Publication of the statement was scheduled for 7 March. Then suddenly it was postponed by the foreign minister on the grounds that, though the cabinet found it acceptable in principle, there were suggestions it wished to make. Van Zeeland nimbly handled the interpellation by reference to Spaak's 1947 statement and by promising further information when technical negotiations were concluded. The delays in obeying the constitutional requirements which in his 17 February message he had suggested were no longer permissible could, it turned out, be tolerated a while longer.

On 10 March the foreign minister indicated what he really wanted: money in lieu of information. The Americans were annoyed to discover that Van Zeeland was again suggesting that they were farther along in industrial developments than they admitted and were holding back. But the key sticking point was a demand for, not a vague promise of financial help for the Belgian atomic program, but an annual contribution of five million dollars. Just what atomic projects this money was to cover was not exactly clear.[25]

The Americans were taken aback by Van Zeeland's sudden turn, by the size of the Belgian financial toll, and by reawakened memory of the tactless terms and excessive demands of the 14 February memorandum which, though superseded by Van Zeeland's March note, seemed still to lurk in the overtones of the more recent communication. They decided not to reply until Assistant Secretary of State Perkins, about to leave for London, had an opportunity to discuss the matter with Ambassador Murphy.

By the time this meeting occurred, however, all Belgium was embroiled in a fiery debate over whether controversial King Leopold III should be allowed to return to his throne. A February plebiscite had approved this return, but by a margin sufficiently narrow as only to add to the controversy. Moreover, the vote was significantly divided by region, with Wallonia strongly against the king while the Flemish voted in his favor.

In May Perkins, Van Zeeland, and Sir Roger Makins, now British deputy under secretary of state for foreign affairs, met in long and inconclusive conversation. The royal affair continued to boil, and on the sixth of June the coalition Eyskens cabinet fell as the Liberal party members withdrew to demonstrate their opposition to their partners' plan to return Leopold. Two days later a homogeneous Social Christian party cabinet was formed in which, however, the Catholic technician Van Zeeland retained his post. In July the king returned to Belgium, whereupon massive demonstrations broke out in Wallonia and in Brussels, led by Leopold's former friend and now most bitter critic, Paul-Henri Spaak.

The possibility of civil warfare loomed. At the beginning of August Leopold grudgingly agreed to turn over his royal prerogative to his son Baudouin and to abdicate when the boy would come of age a few years hence. The cabinet which had in vain endeavored to restore Leopold to the throne also resigned. Van Zeeland, as a key supporter of the monarchy, was invited by Baudouin to form a new ministry. He failed, but he did retain a seat in the next cabinet eventually formed by Joseph Pholien.

Though Van Zeeland had not become prime minister and though he had not succeeded in keeping Leopold on the throne, he had played a key role in preserving the monarchy. He had won the respect of the royal household as much as any politician could without endangering his ability to communicate with the other side, and he had helped spare the nation from further and more serious internecine conflict. If disappointed, he had also reason to feel satisfied. He had stayed in office and would, in fact, survive another cabinet shift in 1952 and serve until Achille van Acker, Spaak, and the Socialists would end their five years in the wilderness by joining with the Liberals to form a triumphant coalition in April 1954.

Anglo-American-Belgian negotiations came to a virtual halt during the middle months of 1950 while the political cauldron boiled in Brussels. Whenever possible, in conversations with Van Zeeland and more regularly with Baron Hervé de Gruben, secretary general of the Belgian ministry of foreign affairs, Perkins and Murphy tried to impress on Brussels officials the real value which the informational assistance proffered did give Belgium. In September Van Zeeland announced that he had reconsidered his position of March and was now convinced that the arrangements posed would bring closer association in the atomic field between Belgium and the two powers. It may have been that the foreign minister was influenced by the presentations made over the summer and that his sharpest domestic political concerns were alleviated by the apparent resolution of the royal affair. These factors no doubt were important, but so too was the invasion of South Korea by North Korean troops in June. Not only did this event underline the Communist menace; it appeared to justify the American emphasis on weapons production. In his message to the Americans and British, Van Zeeland made clear that his posture was shaped by his desire to take into "fullest consideration circumstances as they prevail today."[26] Surely any effort to disrupt uranium shipments would have damaged Belgium's own best interests at that time as well as those of the non-Communist world. Van Zeeland nevertheless reported that he thought the draft communiqué of February could not be considered acceptable. Before it was released he desired clarification on a few points.

These turned out to be related to the earlier oral mention of possible technical aid. An ore reduction plant was contemplated for the Congo. Would the Americans help? The proceeds from such a plant would of course be substantial, and thus its construction would benefit Belgium both scientifically and financially. Security concerns would not be great, for it was intended that the plant reduce ores only to the stage of green salts.

A more sophisticated formula had been derived for the delivery of financial aid—an export tax, the proceeds from which would support a Belgian atomic energy program and related industrial efforts. For the Americans, this was more acceptable than an outright scientific grant, which would have created an awkward precedent for the Combined Development Agency in terms of the other uranium-producing countries. Such a grant, moreover, would have had no relation to the amount of ores produced. The ten-year provision in the 1944 agreement would run until February 1956, but excavation activities were to be accelerated, and it was already expected that Congo reserves might run out before that date. In fact, production was to decline by the end of 1953.[27]

Belgian Ambassador Silvercruys discussed this with Perkins and

Arneson; a week later Van Zeeland, while on route to the United Nations meeting, also visited with Acheson. By then minor wording matters in the communiqué had been worked out and substantial agreement established on an *aide-mémoire* not intended for publication which covered additional points on technical and financial assistance. The size of the increase of the export tax was still in dispute. Van Zeeland anticipated pressure and questions from his parliament, which was to open the next Tuesday. He thought he could postpone statements for a while, but prompt settlement on this one remaining matter would be helpful. Could the *aide-mémoire* simply read that the tax would not exceed 175 francs per kilo, with it being understood that the exact amount would not be determined without consultation with the United States? Acheson replied that the tax was a Belgian internal matter; the United States was sympathetic to the low country's need for funds. As for technical matters, there was disagreement within the American government, but this probably could be solved.[28]

All seemed well. In fact, it was not, and Van Zeeland's rush to achieve cloture on the communiqué and *aide-mémoire* turned once more to foot-dragging. He may have felt justified in this shift because the Americans' official response to his earlier démarche was that they could not give a sensible opinion on the export tax without a more detailed description of the proposed Belgian atomic energy program, "although on the face of it, the tax seemed excessive."[29] Moreover, determination of a reasonable amount for the tax required more examination of the technical and economic feasibility of the Congo plant. Perhaps the Belgian thought it not appropriate for another country to pass on what Belgium in its own sovereign will planned to do, even if it was the other country that would be funding the projects. It was evident, too, that in Washington there were those who were determined that acceptance of Van Zeeland's formula of "not to exceed 175 francs per kilo" should not be interpreted as acceptance of the maximum figure, while in Brussels there were advocates of the maximum.

The Export Tax

Van Zeeland's determination to hold out for a high-level tax became evident in November 1950. That month the CPC countries adopted a revised classification guide which permitted release of information on low-power nuclear research reactors. The state department thought the November announcement of this development would provide a golden opportunity to release the communiqué, giving Van Zeeland material for response to any questions in parliament. The Belgian would have none of it unless firm agreement were reached on the financial issue. He found the U.S. failure to give more specific encouragement of the tax increase "unexpected and disappointing" and saw little use of sending an official to Washington to garner information that would be made public to the world the next day.[30] He rhetorically asked how such a minor advance of information would protect Belgium's interests or give her a square deal.

The Americans suspected that Van Zeeland was in a pique because, in October negotiations, the Union Minière had been granted a fifty-cent price increase. This increase resulted from the American decision, stimulated by the Soviet bomb explosion and the Korean War, to expand production of nuclear weapons. The expansion required more fuel, and that could be obtained only from the Congo by asking Sengier to work the Shinkolobwe mine at a faster but less economic pace of operations. The price increase was not intended to augment Union Minière profits but to extinguish the company's increased costs. Sengier told Silvercruys of the price rise, but apparently the latter did not forward this information to Van Zeeland. When the foreign minister eventually learned of it, he was angered that he had been bypassed, saying that the chambers would be most upset were they to learn that the Union Minière got a price increase while the government did not. Van Zeeland was not impressed by the argument that the augmentation only matched world price increases for all metals and simply covered greater production costs.[31]

Van Zeeland did eventually send men to obtain new technical information, but for several months he did nothing to further financial negotiations. The Americans let the matter ride. Why should they search out additional expenses as long as ore supplies were coming in? Sengier, acutely aware of the Soviet threat and perhaps concerned about possible future inmixing of the Belgian government in his lucrative contracts, was doing a good job of accelerating production. Part of the delay on the Belgians' side stemmed from their efforts to organize the Belgian atomic energy program under a Belgian Atomic Energy Commission.

In March 1951, Pierre Ryckmans, chairman of the new commission and former governor general of the Congo, and several other officials called on Perkins and Arneson. Ryckmans' purpose was to resume negotiations and to justify Belgium's demand for increased revenues. These were needed to finance the atomic energy program, Belgium's participation in a European nuclear physics project, social and economic welfare in the Congo, and military defense of the Congo and the Shinkolobwe mine. The last two items in regard to the Congo were entirely new for the agenda and were herrings at which the Americans were inclined to snap but not pursue. Perkins pointed out that though there were joint American-Belgian recommendations for increased defenses about the mine and for building an airfield at Jadotville, in view of the American feeling that Europeans were expecting the United States to finance the defense of Western Europe without making any economic sacrifices themselves, the Belgians should bear the cost of defending their own colony. Ryckmans conceded the point.

The assistant secretary of state further argued that as the United States was gaining no significant commercial use from the uranium and no economic benefit, there was no reason to think that the Congo should receive economic benefit. Ryckmans demurred, suggesting once again that commercial usage was closer than the United States was admitting. Arneson tried to smooth matters over by saying that while there had been greater delay in implementation of Article 9(a) than anticipated when the Tripartite Agreement was written in 1944, that very delay was the justification for the additional financial and other aid the United States was willing to offer. This led Ryckmans to present his bill. The Belgians planned a graphite reactor, a program of general research and development, $1,200,000 for participation in an Italo-Swiss-French cosmotron at Geneva, and the ore-reducing plant now scheduled for Belgium rather than the Congo. The total amounted to $16,000,000 dollars over four years.[32]

Upset also that Ryckmans was now suggesting the necessity of reworking the draft communiqué on which agreement had existed for a year (differences had continued only over the *aide-mémoire* which was to be annexed to the communiqué), the Americans quickly involved the British. At the beginning of April the two powers presented Ryckmans and Silvercruys with their response. The formal memorandum firmly rejected any expansion of the negotiations to include the economic and social welfare of the Congo and the costs of the colony's defense. Perkins' prior arguments were amplified. The great expense undertaken by Western taxpayers in developing a bomb which directly provided Belgium security was being overlooked; suddenly to add the cost of Congo welfare to the cost of providing Belgium security was unfair to these taxpayers. The American Point Four Plan for economic aid to areas such as the Congo was available, and the Belgians were already considering a program for Congo participation in such Point Four aid. As for the defense of the colony, that should be Belgium's contribution to mutual security.

To a point made in an informal Belgian memorandum of 12 March that the price of uranium had not increased commensurately with other world commodity prices, the Americans replied that the suppliers of raw materials such as the Congo had recently made great profits at the expense of importing nations and that the price for uranium had risen from an average of $1.45 in 1945 to $3.40 and in the middle of 1951 would rise to $3.90. The fairest way to determine price for a commodity for which there was no world market was to build up from production costs. If it were estimated that the Congo would produce 20 million pounds over the next four years, the Combined Development Agency would be paying between $68 and $78 million for that material, depending upon the price charged ($3.40 or $3.90). Eighteen million dollars of this would be for the current 90 cent per pound export tax which would benefit the Congo. The remaining $50 to $60 million represented more than production costs and thus a good portion of it too would go to the Belgian government in taxes. Finally, the allies had provided $13 million for the development of the Shinkolobwe facilities, without which expenditures there would be little ore production to tax.[33]

The British and Americans were sure that the price they were paying was fair and that the uranium's "development under present contracts is producing the fair share of benefits that the Congo peoples have a right to expect from their national wealth."[34] They asked the Belgians to take into account matters of equity and political implication from the American and British points of view, the financial assistance they would give the Belgian atomic energy program, and the real value of the technical aid which the two countries would supply. This they had not come by freely but at great expense of money, talent, and effort. How much aid would the allies give the Belgian atomic energy program? They thought a suitable and equitable amount would be $8 million. This should be delivered over a four-year period as an additional payment for material received. Such an arrangement would mean that the Belgian and Congo governments would receive by way of export taxes alone $26 million plus taxes on Union Minière profits.

The two governments' response far from pleased Ryckmans, who complained that it would make presentation to the Belgian parliament difficult. The wartime agreement had to be justified, and this could be done only by demonstrating that there had been substantial tax increases on uranium since the agreement's conception. The accord now had to be justified to the government party as well as to the opposition and, Ryckmans said, Van Zeeland was initiating the present talks in order to assure the accord's continued existence. Ambassador Silvercruys tried to pave the way for further American and British concessions by suggesting that Van Zeeland would no doubt object to the $8 million figure for Belgium's atomic program and to the two governments' unwillingness to increase their aid to include assistance for Congo welfare.[35]

There can be no doubt that the British and Americans felt they were being held up. Request for reworking of a document which had been approved in principle for a year and in exact words for nearly half a year was bad enough, but the sudden injection of demands to pick up Belgian military costs and extensive Congo social welfare expenses seemed either like the biting of the hand that in so many other ways was feeding and defending the Belgians or as a deliberate attempt to rupture negotiations. Ryckmans' poorly veiled reference to this last possibility seemed petty as well as unrealistic. To whom else would the Belgians sell their ore? The Soviet Union? No. France and Norway desired small amounts for research purposes, but not the amounts currently being shipped; moreover, the Flemish population of Belgium had less trust for the French than for the British and Americans.

It may have seemed to his hearers that Van Zeeland was crying "wolf" too frequently. On several occasions the foreign minister had begged the United States and the United Kingdom for agreement on a major statement. When it was granted in more expansive terms than even Spaak had been able to achieve when battling off Communist-inspired interpellations, Van Zeeland declined to use it in parliament. Though he would then again plead the same problem of having to satisfy the political parties, he in the end would avoid making any far-reaching declaration to the chambers. Since the Communist coup in Czechoslovakia, the blockade of Berlin, the invasion of South Korea, and the revival of the Belgian economy with the aid of Marshall Plan funds, the Communist threat in Belgium was minimal. The chief opposition was the Socialist party, and the foreign policy spokesman for that opposition was Spaak. He could inquire about progress in Belgium's development of atomic energy, but he was not likely to cause any insurmountable difficulty.

Van Zeeland's severe pressure to procure a better return for the uranium may have come from within the cabinet. For years Belgium had been subject to criticism for her and Leopold II's alleged exploitation of the Congo. Many Belgians were anxious to justify clearly their stewardship of the colony. Much had been done, yet much remained to be achieved in terms of education, health care, building of infrastructure, and raising of political sophistication. Demands were growing within both Belgium and the colony for far more dynamic programs for beneficent development of the Congo.

Such programs require vast amounts of funds. At a time when the home country still felt constrained by austerity imposed by the ravages of war, the funnelling of Belgian revenues to the Congo rather than into the social and economic rehabilitation of Belgium proper was difficult to advocate. The situation was further complicated by an old regulation which required that half of all Congo minerals be sent to Belgium; this had been waived, to allow massive shipments of uranium and other items from the Congo to the United States and United Kingdom. Obviously the garnering of increased revenues that could fund both the ambitious atomic energy program which the Belgians wished to undertake for the prestige and future commercial prosperity of the motherland and also welfare programs in the Congo would be a political as well as an economic bonanza.

Despite all the other issues, it was resolution of Van Zeeland's political need to show that proceeds from uranium sales were benefitting the Congo that, along with some easily anticipated financial bargaining, brought agreement. This came not promptly but rather evolved as the Americans grew more cognizant of the foreign minister's problem and the Belgians found a way to make the monies proffered in lieu of the technical information envisaged in Article 9(a) of the 1944 agreement appear as financial aid to the Congo.

On the first of June, the Belgian embassy forwarded Van Zeeland's statement of position in response to the firm Anglo-American memorandum of 5 April 1951. At first glance, it appeared to show little movement, as the message expressed regret that the two governments did not see the Belgian viewpoint and stated that postponement of any conclusion would be less harmful than a settlement open to criticism. A closer reading showed avenues for negotiation which were, in fact, underscored in the text of the memorandum.

Van Zeeland wrote that he was disappointed that the United States and United Kingdom were not able

> to take into consideration the Congo aspects of the problem. . . . Any solution ignoring—or seeming to ignore—these interests would be basically unsatisfactory and likely to prejudice the international position of the three Governments concerned. Propaganda hostile to the Western Powers would not fail to take advantage of what would be labeled a demonstration of economic imperialism at the expense of a nonautonomous territory.[36]

The foreign minister pointed out that the Belgian parliament would require a substantial increase in uranium taxation and that financing of the Belgian atom program "is contemplated through *the earmarking by the Colony of part of the resources derived from uranium* for the purpose of scientific research which will contribute to the valorization of the natural wealth vested in the ore." As for the augmentation of the special tax, he argued that applying to uranium the *"same treatment applied to all its* [the Colony's] *other exports"* should not be construed as implying recognition by the two powers of a commitment by those countries to support social and economic advancement programs in the Congo.[37]

The pathway for compromise was open. Ryckmans had already dropped the demand for Congo military defense funds. Now the two governments would be spared any implied responsibility for economic and social development programs in the Congo. All that was asked in return was, first, some formal statement that tax proceeds from the sale of uranium were going to the Congo. These could, in fact, be the sums proffered in lieu of the information originally promised under Article 9(a). These tax receipts could be used to finance a Belgian reactor, as it would be legitimate for the colony to invest "some" of its proceeds in an atomic energy program without which the Congo ore would have had only minimal commercial value. Second, a proper amount of money was needed, and Van Zeeland proceeded to justify an export tax increase, not in terms of the anticipated costs of the Belgian atomic energy program, but in terms of the general structure of export taxes for the Congo.

Van Zeeland carefully asked, not for a specific dollar increase, but for an augmentation of what the two governments had thought possible in April. Though the tax on uranium, fixed at 60 francs per kilo in 1947 and 85 francs per kilo on 1 May 1948, was higher than the tax on other

Congolese products, the taxes on all others had gone up since 1948, while that on uranium had not. He argued the effect of depletion and then commented that any increase in tax would presumably not fall on the taxpayers of the United States and Great Britain but on the shareholders in Union Minière. This was a clever point, for both Americans and Sengier had been urging that the Belgian government not interfere in the commercial contracts between the Combined Development Agency and the Union Minière. If this view were accepted, it would then be difficult for the two governments to argue that, because they would undoubtedly have to increase payments to Sengier's firm to compensate for the revenues lost in increased taxes, they should let this contract matter interfere with the agreement between the governments. The Belgians also emphasized that the British and Americans were still paying between $2.00 and $2.50 more per pound of oxide content for Colorado and Canadian ores than for Congo ores; thus, even with the increased tax, Congo ore remained a bargain. In closing, Van Zeeland reiterated that "in any public release, the increase be given the character of a profit by the Belgian Congo," with the understanding that part of the resources drawn from uranium taxation would finance the Belgian atomic program.[38]

This last provision on wording had import for the American political situation as well as for the Belgian. Expenditures for uranium could be far more easily defended before the Joint Congressional Committee on Atomic Energy and the American AEC on this basis than on the basis that the Belgians wanted "x" number of dollars for their own atomic energy program. Given the reluctance among the joint chiefs of staff and congress about any assistance for the expansion of the British program, there would probably be considerable negative reaction to the idea of the United States actually funding an atomic energy program on the continent or expressing any form of responsibility for such a program, expecially in a small country which twice had been overrun in the past half century.

A major gathering at the department of state scrutinized these points. Present were three persons each from the Belgian and British embassies, the American AEC, and the state department, plus a representative of the American department of defense. Ryckmans explained that the now hoary joint communiqué draft needed rewording to emphasize Congo interests; these were not mentioned in the former version, and thus it could be criticized "as a bad example of colonialism."[39] The additional tax revenues would be paid into the Congo treasury rather than that of Belgium, as the ores were a Congo resource. But the Congo government could legitimately make some of these funds available to Belgium, because atomic research there would enhance the value of the uranium in Congo soil. Sir Christopher Steel for the British wryly pointed out that by such reasoning it could also be argued that American and British research benefitted the Congo since it increased the value of uranium.

Ryckmans explained that all current uranium tax revenues were going to the Congo and that if future revenues exceeded the needs of the Belgian atomic energy program the Congo government could absorb the surplus, for the wants of the region far exceeded available funds. He assured Perkins that the two governments were not committing themselves to a Congo development program and that uranium would simply be sharing the common tax fate of all Congolese products. Some special recompense was due the region that fueled the bombs as, after all, broad sections of the globe, such as British and French colonies, were enjoying the bomb's protection yet were not contributing substantial amounts of radioactive fuel.

Ryckmans wanted publication of a joint communiqué, but not until full agreement had been reached and wording fixed which highlighted the benefit going to the Congo. Arneson wondered whether the AEC would go along with an ostensible contribution to the Congo rather than a payment under the aegis of Article 9(a). Ryckmans responded that, while the United States would consider itself as contributing under Article 9(a), he thought some easy revisions in the communiqué could slide in Congo interests as well. Gordon Dean thought this could be worked out, but he urged no phraseology that would suggest that the United States had agreed to the tax. Congress would go along best if it were told the tax were imposed as a *fait accompli* with which the Americans had no choice but to comply.[40]

The gathering closed in harmony. Only the financial amount and the wording of the new communiqué remained for settlement. A meeting of Ryckmans, Steel, and Gordon Dean and John A. Hall of the AEC took up the dollar figure. Dean had already explored the issue with the AEC. The Belgians initially had asked for $16 million and the AEC had offered $8. Dean's suggestion of the AEC's "final" offer of $12 million must not have come as a surprise to Ryckmans, who thought the response would be acceptable in Brussels. On the assumption of production of 10,000 more tons through the life of the contract, the $12 million could be raised by a tax of about $1.51 per pound, or about 167 francs, an increase of 60 cents per pound and an amount not really much below the 175 franc tax at which Van Zeeland had hinted months before.[41]

The foreign minister agreed to the settlement, much to the surprise of Murphy and the state department negotiators, who feared additional hard-nosed bargaining from Van Zeeland. The United States need for the raw materials was such that its representatives were prepared to pay the full amount ($16 million) initially demanded. As a procedure of negotiation, however, Ambassador Murphy had suggested that the full sum not be offered immediately lest the Belgians regret that they had not asked for more. Murphy later boasted that by that one bit of advice he had saved his government much more than his lifetime salary.[42] At the time, Murphy thought it best not even to raise the question of an eventual renewal of the 1944 agreement upon its expiration in February 1956.

Van Zeeland felt no pressure to make a statement to the chambers, but prompt agreement was reached on a joint communiqué which would be released "if and when" the Belgian government found it desirable.[43] The covering memorandum of understanding signed by Ryckmans, Dean, and Franks stated that, though existing contracts required the seller to absorb all taxes, the two powers would amend those contracts so that the special tax of 60 cents per pound would be borne by the buyers. The right of Belgium to make further tax adjustments was acknowledged. To meet one of Ryckmans' last concerns, assurance was given that once the U.S. and U.K. governments decided to use uranium for commercial purposes the Belgian government would have the right to do so also. This provision in effect wiped out the requirement of Article 9(b) of the 1944 agreement that, should Belgium desire to use ores for commercial uses, she would first have to consult with and obtain agreement from the two governments. Of course, should Belgium move to commercial usages prior to the United States and United Kingdom, Article 9(b) would remain in effect, but this seemed unlikely.

The joint communiqué agreed upon on 13 July 1951 was shorter than that proposed by Acheson in February 1950. The inception of the wartime accord and its anticipated expiration in early 1956 were again explained. A sub-annex contained a statement of the substance of the 1944 agreement, but this time there was no quotation of Articles 9(a) and 9(b), nor even reference to their content. No mention was made of the initial tonnages delivered or of the subsequent ten-year promise to the two powers of the right of first refusal of all uranium and thorium ores produced in the Congo. The communiqué mentioned vaguely the 1944 arrangements regarding eventual use of Congo ores for commercial purposes. Technical assistance arrangements were acknowledged, such as those involving registration of Belgian students at American and British universities, as was the giving of advance notice to Belgium of the distribution of Congo ore shipments between Britain and the United States. The key clause involving the Congo and the tax read as follows.

> 3. Uranium is a Belgian Congo resource, and substantial sums have accrued to the Belgian Congo Government through the medium of the duty and surcharge on the export of uranium ores. In addition thereto, and taking into account the special position accorded Belgium by the 1944 arrangements, the Governments of the United Kingdom and of the United States recognize that there should be a considerable increase in revenue accruing to the Belgian Congo from uranium to support a Belgian atomic energy research program which will enhance the value of this Congo asset. Accordingly, besides duties levied in accordance with existing legislation and in consideration of the circumstances mentioned in paragraph 2 [efforts to prepare for the time when commercial use of atomic energy might be feasible], a supplementary amount, which, if deliveries continue at the anticipated rate during the remainder of the agreement, would produce about $2,500,000.00 per annum, will be paid during this period to the Belgian Congo.[44]

This wording successfully blended the Congo benefit with the monies offered in lieu of additional information under Article 9(a). The exact amount of the tax was not revealed, so that calculation of the amount of ores to be shippped each year remained a problem. The communiqué did not make it obvious that *all* the additional revenues were earmarked for atomic research in the mother country, presumably leaving nothing of the new funds for schools, textbooks, teachers, roads, airports, hospitals, and x-ray machines in the Congo. Even if the $16 million anticipated cost of the projected Belgian atomic energy program were known, there was no indication that the plan was for the total surtax funds to be used for that purpose or that this indeed was an unwritten part of the bargain struck over the manner in which Article 9(a) would be implemented. As it was, Van Zeeland took no haste even to reveal this document in the chambers.

The Americans and British were pleased to have the matter settled and no doubt glad that over the years they had pressed for rapid development of Shinkolobwe production. Because the great preponderance of ores had already been obtained, the additional cost under the new tax would be small in comparison with what it might have been. Sengier no doubt remained satisfied with his contract, as it was clear since the time of the 1948 tax increase that any further increases would be picked up by the buyers, regardless of the clause about taxes being the responsibility of the seller. He had been kept informed, and it had not taken long for his firm to be squared, once the precise tax figure was set. Sengier may not have been so happy about the destination of the proceeds. He had long advocated that they be used to benefit the Congo and had at one time opposed construction of a reactor in Belgium. Sengier may have continued to think that the bending of the backs of native laborers who lived in huts without modern conveniences deserved a more direct return than Fulbright grants to Oak Ridge for students from Liège and shiny atomic research centers in Geneva, Switzerland, or Mol, Belgium.

The new sums the Belgians were to receive were substantial. They could have received $4 million more. At the time they knew they had pressed the Americans hard and that further bargaining might raise ill feelings and possibly jeopardize the excellent relations which carried many other important advantages. It was, moreover, fortunate for them that they did not allow the negotiations to drag on for another eighteen or twenty-four months, for by that time the Americans would have felt less pressing need for the Congolese ores. And a wait of several more years would have brought the Belgians to the period of much lower uranium prices.

Defense of the Congo

Though the focus of the negotiations was on financial affairs, there was another matter which, though impinging on the discussions only once, is of some interest: the military defense and security of the Congo and especially of Shinkolobwe. As early as October 1945 Secretary of State James F. Byrnes, who greatly distrusted the Russians, was warning the American secretaries of war and navy that the Russians were seriously interested in Libya, primarily because a Soviet base there would give access south to the Belgian Congo and its uranium.[45] A year later Groves wrote Byrnes that Ambassador Kirk had been keeping him posted regarding efforts by Commuunist-inspired elements to infiltrate and organize Congolese workers, stirring unrest "which might be deterred through positive action by this country."[46] He therefore urged the sending of more trained observers to the Congo. Perhaps a consular office could be opened at Elisabethville, which could keep the United States informed about affairs in Jadotville, near Shinkolobwe. Groves was of the opinion that the United States should have consular representatives wherever the British or Russians were stationed.

In the spring of 1948 Edmund Gullion, special assistant to Under Secretary of State Robert Lovett, prior to Spaak's visit to America urged Lovett to suggest staff talks between the military men of the two countries regarding defense of the Congo. American plans were not well developed, if they existed at all. The Congo would be a prime objective of airborne operations in wartime, but the United States was not in an advantageous position to undertake them. The British had bases closer to the Congo; it was suggested talks might be coordinated with them. Their chief of staff, Viscount Montgomery, had already visited the Congo.[47]

The joint military discussions apparently never got underway. That November, as the terms of the North Atlantic Pact were being discussed, Spaak raised with Ambassador Kirk the possibility of the Belgian Congo and its uranium being included within the area to be covered by the treaty. He could understand why inclusion of North Africa was being contemplated, as the allies needed bases there. But, then, should not the Congo also be protected? He was informed that the Americans preferred that no African territories at all be included in the pact. Were this the decision, the problems Spaak anticipated in the chambers would be lessened. In any case, the department of state reported it had approached U.S. military authoriities, who attached the "highest importance" to the Congo's inviolability but could give no promises.[48] Subsequently, the Belgians were invited to convey to Washington any specific concerns about Congo security.

Meanwhile Brussels officials were considering building an airfield at Jadotville, thirty kilometers from the mines. Kirk reported that plans called for a a landing strip which would not be adequate for combat conditions. He wondered if the department could encourage construction of the airfield and perhaps provide financial aid by funnelling money through the Union Minière.[49]

At the beginning of 1949 Secretary of Defense James Forrestal asked about Belgian provisions for the security of the mines and transport of ores to shipping ports. Kirk reported only that the *Force Publique* in the Congo numbered barely fifteen to seventeen thousand men, which the embassy considered wholly inadequate. A native police force of just fifteen men was stationed at the mine, headed by a newly appointed white man. The nearest *Force Publique* garrison was at Jadotville.[50]

As Kirk thought further on the situation, he was far from reassured. The ambassador believed the Russians had the capacity to fly troop carriers from satellite bases to Katanga and return without fueling. It would also be easy to damage the mines. The Russians had to be credited with at least as much foresight as the Japanese displayed in attacking Pearl Harbor. They had pioneered development of paratroops and had "almost unlimited trained manpower." The Soviets might well decide to fly in a force adequate to wreck the mine. Given the weakness of the *Force Publique*, "if Russian transport planes are permitted to reach the mine the paratroop operation would be a child's play."[51]

Reports of expanding Communist activity in the Congo further excited Kirk's anxiety. The Central Intelligence Agency was also concerned. Twice in the previous eighteen months the state department had responded negatively to CIA inquiries about the value of sending a CIA agent to Leopoldville, capital of the colony, or to Elisabethville in Katanga province. The state department had not favored CIA involvement because of "chronic Belgian suspicion of our motives in the Congo, as well as official ill-feeling"[52] that stemmed from the misbehavior of an agent in Leopoldville during the war, who Governor Ryckmans eventually requested be removed. Now, however, British agents were being received in a reasonably friendly manner, so the CIA was given the green light to send its own man.[53]

The closing months of 1949 found the United States and Belgium in negotiations leading to the Mutual Defense Assistance Agreement formally signed on 27 January 1950 and entering into effect on 30 March. A key intent of the Americans' Mutual Defense Assistance Program (MDAP) was to enable allies to build their defenses in Europe. In Belgium's case, the American military saw the need for keeping the Congo safe.

When Belgian representatives explained that they wished to ship Belgian metropolitan forces to the Congo, the Americans quickly made a gentleman's agreement that conditions could be worked out whereby Belgium could in the future send Mutual Defense Assistance Program equipment to the colony without having to work out any special agreements at the time. At first it was thought an exchange of letters or a memorandum of understanding could be arranged for signing simultaneously with the bilateral treaty.The Belgians submitted a draft regarding the transfer of MDAP material to the Congo for training of troops there and for defense of bases in the colony. The department of state found little problem with the transfer of material but did feel that any agreement regarding defense of Congo bases required a separate and specific request. Moreover, the negotiators involved began to worry that other European allies, were they to learn of the special arrangement, might also ask to use MDAP equipment to defend their colonies.[54] This could lead to a weakening of the defense of Europe and thus of American security, defeating the chief purpose of the Mutual Defense Assistance Program. It could also provoke congressional opposition to the funding of that program through the implication of American responsibility for the military defense of colonies. So it was that any exchange of letters or special agreement for defense of Congo bases was postponed until after the signing of the bilateral Mutual Defense Assistance Agreement.

The delay proved longer than anticipated. American concern for the vulnerability of the Congo continued, and information was gathered about the region; yet little concrete was done. In November 1950 Ambassador Murphy visited the Congo for the explicit purpose of evaluating defense conditions. Though he found the inaccessibility and immense size of the area to be an asset as far as defense was concerned, he urged that:

> thorough-going steps should be taken to protect this important source [of] uranium and other minerals, especially cobalt. Present measures are definitely not adequate. It is believed that such measures can be taken at relatively small cost.[55]

Inquiring into Belgian defenses of the Congo was a ticklish matter for the Americans. Sengier himself had made clear since 1944 that for business reasons he did not relish having foreign visitors snooping about. This feeling was widely shared throughout Belgian official circles, as American criticism of the practice of colonialism incurred irritation and even anger. As Murphy wrote Acheson at the end of 1949,

> Belgian officials, both in the Congo and in Brussels, who deal with colonial affairs have an almost morbid hypersensitivity with respect to any United States interest in the Congo. There is an invalid

but nevertheless deeply seated opinion in these governmental circles that the United States has egged on irresponsible members of Committee IV of the General Assembly and the Trusteeship Council of the UN to an undue interest in affairs which many Belgians feel are limited strictly to Belgians in the Congo and appropriate business and government circles in Brussels. There is no doubt but that Mr. Wigny [the minister of colonies] shares this widespread opinion.[56]

The Americans had to tread lightly, and the path chosen was that of a joint Belgo-American Commission on Congo Defense which began work in late summer 1950. The Commission concluded that the Shinkolobwe mine was of capital importance to the free world, that Belgium should take all necessary measures for its security, and that the *Force Publique* was capable of coping with neither internal dangers nor naval or aerial attack. Recommendations included asking the North Atlantic Treaty Organization to evaluate the probabilities of attack, to reinforce security dispositions in the province of Katanga, and to build an adequate military airfield at Jadotville. The Commission also wanted personnel and matériel procured which would meet the recommendations of a committee of experts, steps taken for the creation of a liaison mission until there was a unified command of Belgian metropolitan and colonial forces, and a study made of anti-aircraft defenses for Katanga and all the lower Congo River.[57]

While the Commission was working, the American joint chiefs began formulating their own recommendations, drawing upon embassy reports; those of the Belgo-American Commission, which sent a binational survey team to the Congo at the end of 1950; and the views of General Handy, who also visited the region. The joint chiefs did not take quite such an alarmed view as did the Commission. Land or sea attack on the Congo seemed improbable; some risk of air attack would have to be run, for construction and maintenance of an appropriate defense system was not feasible. The greatest danger lay in a large-scale native uprising or in dissatisfaction among the mine workers that would lead to disruption of production. There was also a possibility that the Soviets might overrun Western Europe and that a collaborationist Belgian government might cut off ore supplies. The joint chiefs therefore recommended that the Belgian government increase defenses of the mining area against internal disorders and external attacks by posting more forces and strengthening intelligence activities. At least one shipping route from the mine to a port should be well protected. They wished that Brussels officials be asked to institute programs to maintain high morale among the Congo natives to counteract Communist activities. Also, the director of the CIA should initiate plans for "covert sabotage," presumably of Communist leaders and propaganda, to improve security of the Congo and especially of the Shinkolobwe mine area.[58]

The contrast between the two sets of recommendations is striking. The group which included Belgian officials focused on the danger of external attack and requested financial aid for specific projects, such as massing of matériel and the building of an airfield at Jadotville—something Kirk also favored. The American joint chiefs focused on the internal danger and native morale. Each group saw different dangers and emphasized what the other country should be doing. It was to the recommendations of the Belgo-American Commission that Ryckmans referred in March 1951, when arguing for increased revenues from uranium.[59]

The Americans, however, did not think that a large airfield was required at Jadotville and soon so notified Brussels. As for the Belgian idea of having NATO apprise the possibility of attack on the Congo, again the Americans considered it undesirable. They did not want the newly formed alliance weakened by implying to any of its members that they might have to take responsibility for defending Belgian rule in the Congo. Aware of Belgian sensitivities and of active measures already undertaken to improve native morale, Murphy arranged that anything said would be to praise and recognize these actions, rather than to prod.

The Belgians were told they would get a reasonable amount of military aid beyond what they themselves could afford for the Congo. A list of equipment needs was screened by the American Military Assistance Advisory Group in Brussels and referred to the joint chiefs for final approval. Steps were taken to station a CIA agent in the Congo to further counter-sabotage. At the time agreement was finally reached on the export tax, the Belgian ministry of colonies was studying the issue of increasing civilian and military intelligence forces in the Congo, a new battalion of native troops had been created to protect the Shinkolobwe mine and its electricity supply, anti-aircraft needs were being estimated, possible unified command was being examined, as was the stationing of more European troops in the Congo, an additional transport squadron had been ordered to the colony, and progress was being made on construction of an airbase at Kamina.[60]

Thus by the middle of 1951 agreement had been reached on matters both economic and military. The Belgians had successfully obtained some American military aid in addition to a good surtax figure, while the Americans had successfully avoided taking a portion of responsibility for the economy or social welfare of the Congo and had assumed only limited obligations regarding its defense. The British and Americans had their ore, Van Zeeland a stronger political position, and Belgium more money, more information, and an ambitious funded atomic energy program. The Congo would get a few more troops.

It is surprising that the Americans, so prompt under Groves' leadership in establishing diplomatic control of the uranium, did not have even one CIA agent active in the Congo in 1951. This discrepancy reflected not only awareness of Belgian sensibilities but also the slowness with which the Americans became awakened to the full burden of their new superpower position, which they owed to such great extent to the possession of the Congo uranium-fueled bomb.

Devising a program for the protection of the Congo, especially Katanga, and finding the funds to implement it remained problems for the Belgians. A special interdepartmental study group was established in Brussels, and a technical mission sent to the colony. Though neither group was scheduled to report before the fall of 1952, in July of that year Ambassador Silvercruys contacted Perkins and Arneson to report that preliminary indications were that it would be extremely difficult, if not impossible, for Brussels to provide the military equipment needed. He referred to the projections made at the time of the negotiation of the bilateral Mutual Defense Assistance Agreement of further understandings on the supplying of materials to Belgium for use in the Congo. Would the United States be willing to consult on the matter?[61]

Consultation was easy, Perkins said, but provision of funds would not be. Congress had cut military appropriations, and the United States simply could not do what it wanted to do even in Europe, much less elsewhere. Arneson pointed out that in the past there had been differences between what the Belgians viewed as necessary material and the opinion of the American joint chiefs of staff. The Belgian request for assistance in 1950 had totaled about $25 million because it included substantial amounts of anti-aircraft warning and defense equipment. The joint chiefs, considering the possibility of airborne attack on Katanga as remote, cut the request list to $7 million. They then made that equipment available as reimbursable assistance under Section 408 of the Mutual Defense Act of 1949. The revised military assistance list had

223

been submitted to Belgium. No reply had been received, perhaps because of the death of Franz Leemands, chairman of the Belgo-American Commission. What Arneson wanted was a "barebone, realistic assessment of the equipment needed for effective ground defense of the Katanga."[62]

As Perkins commented, receipt of overlapping and divergent recommendations from different sources was confusing and could delay any delivery of aid. It was clear, however, that American aid would not be as extensive as the Belgians desired for three reasons: different conception of the nature of the danger, U.S. unwillingness to take major responsibility for protecting Katanga from internal threats as compared to external attack, and lack of funds. The declining importance of projected uranium shipments may have been a factor in reducing American interest. Yet this was counterbalanced by the increasing need and relatively short supply of cobalt, a major Congo export necessary for the making of the high-density steel essential for armament-piercing projectiles. By August of 1952 Acheson was authorizing security surveys—eventually made by American representatives—of cobalt operations at Jadotville as well as of the uranium mine at Shinkolobwe.[63]

In American eyes, the danger of external attack continued to be remote, while the problem of internal security seemed to mount. Roger Mellen Bearce, vice-consul at Leopoldville, told of deterioration of rapport between the native and European populations, increased surliness and poorer job performance by the workers, and the like. Fears of "an organized campaign of induced disaffection" grew.[64] More investigative studies were made, including the visit of a Belgian parliamentary commission. In 1953 Senator Bourke Hickenlooper led a congressional delegation to Africa and visited Katanga, presumably as a goodwill gesture. Belgian reaction to the visit was negative, as some persons thought it meddlesome and likely to focus increased attention on the sensitive Shinkolobwe area, which, however, Hickenlooper was not allowed to visit. In short, events were turning to the path leading to the rapid declaration of Congo independence in 1960 and the chaos which ensued. The external threat envisioned by a few in 1950 was clearly being replaced by internal problems. The Pentagon's reluctance to become more deeply involved was evident. Yet in Brussels there were individuals who anticipated that, should events get out of hand, the United States because of its reliance on Congo cobalt and uranium might lend support, or at least be sympathetic, to special defensive actions in Katanga.[65]

9 · Atoms for Peace, Atoms for War

One wonders what the emotions of Charlie Steen were when his test drill finally broke in the arid, brown and red Lisbon Valley in Southern Utah. Anger, resentment, frustration, discouragement are all good possibilities, for he had been prospecting several years without success. Steen had hoped to locate yellow carnotite, the ore in which uranium was expected to be found, yet the core his drill brought up was black. He threw it in his jeep anyway. Too broke even to own a working geiger counter, he did not know until he encountered a friend with a healthy counter that his sample was of radioactive pitchblende, containing "a bonanza that would change the course of United States geology."[1]

The year was 1952. Only a few months earlier in 1951 Paddy Martinez, a Navajo Indian, noticed that uranium samples at a trading post were similar to rocks he had seen along the Sante Fe railroad right-of-way near Grants, New Mexico. Further explorations were to prove these to be a major and unusually accessible uranium find.

Nineteen fifty-two was the year that South African uranium began arriving in the United States in appreciable quantities. It was also the year that the new, more efficient Redox ore-processing procedure came on line, thus significantly improving yields. Increased deliveries were being received from the new processing plant in Canada. Additional supplies from Rum Jungle and Radium Hill in Australia were anticipated, though the huge resources of the Algoma region of Ontario, Canada, had yet to be recognized. In short, by 1952 the Anglo-American fuel shortage was over. The American Atomic Energy Commission began limiting its purchases in 1950, and by fall of 1951 it recognized that by the end of that year ore deliveries would exceed requirements. As the official historians of the AEC have commented, "At last, availability of raw materials would no longer be a limiting factor in the nation's atomic energy effort."[2]

This did not mean that the United States and Britain would abandon their efforts to assure themselves of a sufficient uranium supply. This they would continue to do for a number of more years. After all, in 1952 the uranium produced in the United States amounted to only 25 percent of the nation's total procurement. But the future growth of this percentage was assured. Plans were made to cease all direct purchases by 1962; the failure of commercial demand by that date to absorb the uranium industry's production led to two different AEC "stretch-out" purchasing programs which finally concluded on 31 December 1970. By that time the AEC had a surplus of 50,000 tons of U_3O_8 on hand; three years later the Commission was attempting to determine an appropriate means of disposing of it. From 1947 to 1970 the United States AEC purchased 174,500 tons of U_3O_8 from domestic sources, nearly 55 percent of its total acquisitions for the period, 73,800 tons from Canada, and only 67,600 tons from overseas.[3] In short, by the early 1950s some of the old needs were dissipating; at the same time the efficacy of preemption efforts would appear increasingly questionable and costly in relation to benefits received and information bartered.

Revision of the Atomic Energy Act

If 1952 marked a new stage in the search for rare ores and a new stage in the nuclear weapons race with the detonation of the American hydrogen bomb, it also witnessed a strengthening of the movement to do something about the 1946 Atomic Energy Act. The AEC had been encouraging private industry to assist in industrial development, yet any meaningful participation was precluded by the terms of the McMahon Act. The government's monopoly of materials and information was described as "an island of socialism."[4] If the industrialists were to become truly involved and give the program the boost that private enterprise could contribute, then the industrialists needed incentives.

The complications the act brought to relations with other countries had long been evident to the state department and to the AEC. The supplying of funds in lieu of information had briefly placated Belgian feelings, yet each forward step in atomic energy development announced in the United States would mean further embarrassment. Belgian Ambassador Silvercruys naturally reported home stray comments, such as AEC member Eugene Zuckert's statement that nuclear production of electricity was closer than had earlier been expected. Relations with the British had been marred, and negotiations with other countries hindered. Dean's success in obtaining a softening of regulations sufficient to give the Canadians needed information in 1951 did not extinguish the problem but rather highlighted it and the probable virtue of even greater changes in the regulations. The U.S. military, for its part, wished some alteration in Section 10 of the act so that more information could be given to North Atlantic Treaty Organization allies regarding the effectiveness of the weapons. This would facilitate better planning. General Omar Bradley of the joint chiefs of staff pushed for change and was supported by the state department.[5]

In any case, other nations were moving ahead with their own atomic programs. Each desired information from the United States, and American reluctance either to volunteer help or to respond favorably to direct requests for information did not win friends. Allies were expending monies that could have been directed to other social and defense goals; they were beginning to talk among themselves since the Americans would not talk with them about atomic research. It was almost inevitable that these nations, intrigued by what appeared to be the revolutionizing power source of the future, would wish to develop their own capabilities. Had the Americans shown more willingness to work with them, they might well have relied—as did the Belgians for so long—on bilateral information exchange with the United States and Great Britain to further their own programs. Without such cooperation, the nations naturally turned toward each other. It was a considered policy of the Americans to further European unity. In their refusal to share information and material, they managed to encourage such unity in perhaps the one area in which they were not eager to see it occur.

On several occasions before assuming the presidency in 1953, General Dwight D. Eisenhower had spoken in favor of better cooperative relations with the British on information sharing. His record in favor of international cooperation was an established one, soundly substantiated by his performance as supreme commander of allied forces in Europe during World War II. His speech to the United Nations General Assembly on 8 December 1953 was therefore well based on previous positions and philosophy. Yet it took many of its hearers by surprise and had an electric effect.

The president had, in fact, been searching for some months both for a means of reopening the atom control issue and for a way to give his United Nations appearance significance. He wrote in his memoirs that during 1953

I began to search around for any kind of an idea that could bring the world to look at the atom problem in a broad and intelligent way and still escape the impasse to action created by Russian intransigence in the matter of mutual or neutral inspection of resources. I wanted, additionally, to give our people and the world some faint idea of the distance already travelled by this new science—but to do it in such a way as not to create new alarm.[6]

In September Eisenhower raised the notion of the U.S. and the Soviets both turning over the the United Nations "x kilograms" of fissionable material to be used for peaceful purposes. Lewis Strauss, now the new chairman of the Atomic Energy Commission, would suggest the concept of storing the uranium in solution in a diluted manner to prevent theft. The proposed speech was worked and reworked and then subjected to further minor changes after Eisenhower discussed it with French Premier Joseph Laniel and, in more detail, with British Prime Minister Churchill at an allied conference in Bermuda immediately preceding Eisenhower's United Nations appearance.[7]

Corbin Allardice, at the time executive director of the Joint Congressional Committee on Atomic Energy and charged with maintaining close relations with the AEC, later commented that Eisenhower hoped the creation of a pool of fissionable materials from which all nations would draw for peaceful purposes "would create a demand large enough to drain away material that might otherwise be used for weapons. To some extent, this concept was based on the faulty premise that there was a shortage of raw material."[8]

Eisenhower in his diary indicated several reasons for the speech. One was to get the U.S.S.R. to work with the United States on some atomic plan, another to interest the small nations of the world in the uses to which the "limited supply of raw material" would be put.[9] He wanted the American population to believe its expenditures were not solely for destruction. And another reason was that if the U.S.S.R. did participate,

the United States could unquestionably afford to reduce its atomic stockpile by two or three times the amount the Russians might contribute to the United Nations agency, and still improve our relative position in the cold war and even in the outbreak of war.[10]

At Bermuda, the purpose Eisenhower emphasized to Churchill and Laniel was reduction of the Soviet stockpile.[11]

Thus the concept of limited resources of uranium lingered. Strauss would write in his memoirs that uranium "was still an excessively rare commodity and the key to control of atomic energy. The great reserves of uranium in the United States were scarcely suspected."[12] To whatever extent that concept influenced Eisenhower's thinking, the race for uranium carried significance. If Britain and the United States thought they had uranium to spare, they believed the U.S.S.R. did not.

The president's "Atoms for Peace" initiative provided the stimulus for revision of the international aspects of the 1946 Atomic Energy Act. Within the AEC there was strong support also for revision of the clauses restricting collaboration with private industry. In due time the AEC prepared two bills for revision of the act. The state department urged that the legislation be drafted in a manner which would allow the United States to deal with other countries to assure continuance and enhancement of uranium shipments; this was needed if the Tripartite Agreement with Belgium were to be renewed. Properly phrased, the amendment could help bind allies closer and influence neutrals to be more cooperative. Moreover, the United States would look bad in the eyes of the world if some other power were to precede it in developing and sharing useful domestic nuclear power.[13]

A number of legislators supportive of international cooperation feared for the public interest if private industry were to become more deeply involved in the production of atomic power in the form of electricity. On the other hand, several supporters of private industry involvement viewed international cooperation dimly, perhaps reluctant to lose business advantages which could accrue to American firms. The separate proposals for revision of the act thus faced strong possibilities of legislative failure. The solution, as proposed by Allardice, was the combining of the international and domestic provisions in the same proposal.

Eventually the new act passed; Eisenhower signed it into law on 30 August 1954. The atom was freed from government monopoly; the path to commercial development lay open. So too did the way for better cooperation with other nations. In November of that same year Eisenhower allocated to the Atoms for Peace program the first 100 kilograms of American-produced U-235. The government retained the right to restrict possession, transfer, or use of key materials for reasons of public health or national security. It could control facilities for production of sensitive materials and could determine what knowledge would be kept secret and under what circumstances it could be disseminated. But dissemination of information was permitted, with the now customary procedure that proposed cooperative international agreements on atomic matters had to lie before the Joint Committee thirty days while congress was in session.

A new era in Western atomic relations was about to begin, based on fuller informational exchange and international cooperation between the United States and Great Britain and between these two and other nations than had existed at any point in the previous decade. It would also become based on the premise, not of the universal scarcity of uranium, as had the period of preemption of the rare ores, but of the basic availibility of uranium. The desire to avoid proliferation of weapons would continue, but the emphasis would switch first to limitation of production facilities such as gas diffusion plants and then to holding secret techniques for reducing the size of bombs and to limitations on delivery systems—bombers and missiles. After all, Britain had the bomb, as did Russia; the French were beginning to develop significant domestic uranium resources and had two successful reactors in operation. Norway and the Netherlands jointly completed their JEEP reactor in 1951, using Norwegian heavy water and Belgian uranium purchased by the Dutch as early as 1939 for such possible use. Sweden was completing her first reactor, and the Danes were undertaking significant research. So too were the Italians and the Swiss, in whose country the Central European Research Laboratory was developing at Geneva. Brazil, India, and Australia were also moving forward. Belgium was behind several of her European neighbors, but the funds provided by the uranium agreement of 1951 enabled the founding, the following year, of the Study Center for Application of Nuclear Energy (later renamed the Nuclear Studies Center) at Mol.[14] Despite the dikes built by the Americans to retain their secrets, knowledge of nuclear processes was spreading.

Negotiations with other countries of course did continue. Yet the American Atomic Energy Commissioners found themselves more frequently caught between their old purchase-or-preempt-at-any-price policies and the recognition that uranium and thorium were becoming increasingly available, that supplies were greater than the need, and that the federal budget required trimming. This was demonstrated in the continuing negotiations with India, Brazil, and South Africa. Those negotiations also reflect how, although the Americans and British maintained their partnership in the years after the war, the lead in raw materials dealings had gradually passed to the Americans by the early 1950s.

Britain, confronted by economic and imperial problems, simply did not need or want either the extensive supplies or the high expenditures associated with the massive and occasionally preemptive purchases the Americans were willing to arrange. Even where the British influence was most strong, as in India and South Africa, by the beginning of the 1950s the British were for the most part either going along with the Americans' pace or almost relinquishing the field to the insistent American negotiators. To a considerable extent this was appropriate, as Britain no longer had to pay half the cost of the materials but only for the amounts she used. In the instance of thorium, she had virtually no use for the material and increasingly less interest in attempting a preemptive cornering of the market. Yet by the peculiarities of her arrangements with the United States even Britain's ability to share information with third countries in return for raw materials was fettered by laws in the making of which the United Kingdom had no voice. If, as in the case of Norway, the British would be unable to be of help because of the U.S. position, then the Americans might as well set the parameters for the technical exchanges at the beginning. This is not to say that the British did not provide technical assistance to other countries. They did, but not in areas where the information was crucial to the development of fission energy.

India and Brazil: Foreign Aid and the Purchase of Unwanted Materials

Prior to 1954 the McMahon Act of 1946 was of course the key piece of legislation which affected the bargaining of technical information for rare earths. A second was the Mutual Assistance Control Act of 26 October 1951; it was otherwise known as the Battle Act, named for its chief sponsor, Representative Laurie Battle of Alabama. For a number of months prior to the passage of this act, the state department had been successfully negotiating informal agreements with various nations whereby they promised to control export of strategic materials and equipment to other countries or at least ship only to those nations approved by the United States. Battle's legislation took the issue farther, as it required that all United States financial, economic, or military aid to another country automatically be terminated if that country were found to be trading embargoed items to the U.S.S.R., its satellites, and Communist China. Two lists of materials were maintained. List A named the most highly strategic items, including uranium and thorium nitrate. List B carried items of somewhat lesser importance such as beryl, for which, at least prior to the Battle Act, embargo expectations were not so complete.

On 17 July 1953 the Polish freighter *SS Mickiewicz* pulled out of Bombay harbor, carrying in her hold 2,248 pounds of thorium nitrate. This cargo, fully paid for, had been sold by the government of India owned firm of Indian Rare Earths Limited and was destined for Taku Bar, Communist China, by way of Colombo, Ceylon. Secretary of State John Foster Dulles immediately urged his ambassador in New Delhi, George V. Allen, to take the "extremely grave" matter up with the Indian government.[15] A great amount of American aid flowed to the newly independent state, but the Battle Act allowed no executive discretion which could blink at such a shipment of listed materials.

When confronted, Indian Foreign Minister Raton K. Nehru admitted that he had not realized the full implications of the Battle Act. The shipment was small, too small for atomic weapon development, and intended to help open Communist markets to Indian goods. The prime minister could of course order off-loading at Ceylon, but the foreign minister thought this would cause difficulties. Should it be done, it would reveal destinational discrimination in exports, something contrary to his government's announced policy. Moreover, it would reveal political strings to United States aid.

This last was a telling point, for the Communist press had for some while been asserting that the United States was forcing aid on India in order to entangle that country in various attached commitments. The proud young country had, for its part, carefully avoided specifically requesting American aid. To do so would be to admit that India could not survive on her own, once relieved of colonial status. To spare the Indians such embarrassment, the department of state had therefore informally ascertained the nature of India's needs and then proffered aid. This the Indians could accept without losing face or appearing a vassal of any of the great powers. The state department, concerned that economic conditions in India made the country ripe for Communist subversion, strove hard to convince congress to make available the sums thought minimally necessary. The Indian government had been informed of the Battle Act on 12 December 1951, and the India-United States aid agreement had been signed on 5 January 1952. Thus, whatever limitations did exist on American aid were known by the Indians before the aid was accepted and, as far as the Americans were concerned, did not consist of "strings." The latter term, they argued, could be used only if the United States asked for some unanticipated political action after the aid were granted, with the implied threat that the aid would be pulled back if the recipient did not comply to this new demand.[16]

Cessation of American aid would most likely end any hope of India realizing the goals of its five-year plan, achievement of which the state department saw as an economic and political necessity. Moreover, cancellation would play directly into the Communist propaganda plan. Any attempt to restart aid after a brief stoppage was likely to fail over congressional requirements for specific assurances that there be no repetition of the thorium shipping incident. Because the department of state had avoided requiring specific commitments from the Indians in the earlier negotiations because of their sensitivities, it was improbable that the proud New Delhi statesmen could be brought to even more binding wording after aid had been withdrawn.

Two American assistant secretaries of state thought there was a way around the dilemma. Though Prime Minister Jawaharlal Nehru had himself signed the export license, he claimed this was done as a routine act while approving a number of documents. Lower-level officials insisted they did not understand the implications of the shipment for the aid program. Therefore the argument could be made that the Indian government had not "knowingly permitted" the shipment—to use the phrase provided by the Battle Act and a restricted definition of "knowingly." The recommendation was that Harold Stassen, administrator of the Battle Act, should so rule, thus permitting no disruption of aid. Meanwhile, the Indians were being well informed of the strict wording of the legislation.[17]

Inadvertent or planned, the shipment was an effective way to get the Americans to the bargaining table. The United Kingdom and the United States had earlier negotiated for purchase of thorium nitrate. The talks had failed on the issue of price. Now Dulles was faced with either having to condone possible successively larger shipments of thorium to China, perhaps for transshipment to Russia, or terminate American aid to India. Neither course was in the best interests of the United States. Nor would either favorably impress congress. Termination of aid would fuel the arguments of those congressmen critical of American aid from either the conservative or liberal viewpoint, and extended condoning of illegal shipments could cause an even greater executive-congress row.

Ambassador Allen saw the light quickly and urged Washington to resume negotiations. How could the United States hold the position of denying the right of a third country to trade with iron-curtain nations unless the U.S. agreed to purchase the third nation's goods itself? The Indian nitrate plant's output could not exceed a value of a half million dollars per year. That was less than one percent of the some $75 to $90 millions of aid destined for India. Why not purchase the thorium nitrate, support the nascent processing industry in India, save the aid program, and please aid critics with a specific example of "trade not aid?"[18]

By the beginning of September Dulles had accepted the recommendations of his assistants on dealing with India. Stassen was contacted, and a price of $2.75 per pound of thorium nitrate was suggested to Allen, for this was a price informally put forward by Dr. Shanti S. Bhatnagar of the Indian AEC as likely to be acceptable. Bhatnagar soon raised his suggestion to $3.05, for that was the price India had paid the United States for thorium nitrate before it had its own processing plant.[19]

The American AEC was prepared to pay only $2.20 per pound for thorium nitrate. Supplies were now plentiful, and use for atomic energy purposes still unsure. The AEC was buying the nitrate elsewhere, especially in South Africa, and acquiescence to a higher price for India would cause price structure difficulties. Moreover, the AEC saw any premium over its base price as a form of foreign aid and thus not really a responsibility of the AEC. Allen was authorized to make an offer as high as $3.20 per pound if necessary for a contract for two years or 135 tons, whichever came first. The extra $242,000 over the $2.20 price would be made available either through grants funded by sale of surplus U.S. equipment in India, by the purchase by India of equipment in the United States, or by sale of an option on Indian rare earth products. In addition, Allen negotiated for an oral statement from Nehru about shipment of strategic commodities.[20]

In October, the government of India raised its price request from $3.05 to $3.25 and then to $3.50 per pound. On 16 January 1954 the state department acquiesced to the $3.50 per pound rate "to stop further haggling on price which might also permit GOI [Government of India] to open discussion on other points." The contract amount was also increased to 230 tons of thorium nitrate. Dulles told Allen that

Conclusion of contract and fact that potential explosiveness total situation appears [to] be increasing with passage of time point to advisability [of] not pressing further discussions on additional Battle Act assurances. . . . Believe that, in this situation, US may well attain its objectives by putting matter squarely in hands of GOI with implication moral responsibility to effect control over Battle Act items.[21]

The agreement reached was for a term of eighteen months from 1 April 1954 or until 230 long tons of thorium nitrate were delivered; India was permitted to sell small amounts of thorium nitrate for commercial purposes to specifically named nations.[22]

The affair demonstrated the increasing questionableness of the practice of preemptive buying of thorium. Its actual present or future use or need was dubious, as was the possibility of restricting all Communist forces from access to world supplies. The continued refusal of the AEC to go above the $2.20 per pound price for nitrate which scuttled the initial talks is significant. In short, the issue was no longer a direct one of free world security but one of foreign aid to India, which aid needed to be extended one way or another because it helped to stabilize the new nation's economy, thus indirectly aiding Western security.

Prime Minister Nehru may not have knowingly intended to jeopardize all American aid to India when he authorized the shipment of thorium nitrate to China, but he clearly was using it as a way of forcing the Americans to pay more money than they wished for material they really did not want. The decision to pay the price was basically a foreign aid and foreign relations issue, not one of procurement of a rare ore. The negotiations therefore differed from those of earlier years. It was not the Americans who took the initiative or controlled the pace. The patience and flexibility shown by the Dulles state department was greater than granted by its critics at the time. Some irritability at being so maneuvered is understandable, and it is not surprising that a few officials did discuss the subject of "diminishing returns" in showing such forebearance and generosity.[23] If Nehru were far from subtle in this instance, the Americans had also been heavy-handed in some of their dealings. The end result of the thorium discussions still served the vital interests of both countries. Yet one can wonder what price would have been paid, both in dollars and in terms of Western security, had the Belgians taken an attitude in 1944 or 1947 similar to that of India in 1953.

Though the Americans gave way on the thorium issue, the case was different when the 1 October 1950 beryl contract with India came up for renewal. Decision had to be reached one year in advance. Shipments had averaged about $150,000 in value per year, and the AEC saw no need for a five-year renewal as the terms of the contract stipulated. American requirements were such that the Commission could countenance only a one-year renewal. The state department, however, desired renewal for the full five-year period because, should new needs develop in the future, another contract would be difficult to negotiate; moreover, the beryl contract presented another case of "trade not aid" and held the potential for another breach with the Battle Act, for beryl was on its Category I-B list of materials of primary strategic significance.[24]

In deference to the views of the AEC, the state department approached India with the suggestion of a one-year extension of the existing accord. At the same time, it presented the matter formally to both the AEC and Stassen of the Foreign Operations Administration; again the goal was to have the Operations Coordinating Board arrange to have the cost of a possible five-year contract assigned among various government agencies. Once more it was a case of agreeing to buy what was no longer rare or badly needed in order to maintain good foreign relations, rather than relying upon good foreign relations to facilitate reasonable sale of a rare and highly desired commodity.

The Operations Coordinating Board decided in favor of a five-year renewal, but the AEC on 24 November reopened the matter by coming out for only a one-year renewal, with the option of an additional one-year term. The debate among the American officials had continued because the Indian government had allowed a delay of three months in the consideration of the matter. The differences stimulated considerable discussion at the OCB meeting of 15 December. Representatives from the state department pointed out the serious political repercussions that might follow if the United States refused to purchase beryl but kept it on the Battle Act lists of restricted items of trade. The AEC at first argued that, though it had no current need for the mineral, because it was used in aircraft construction and in reactors, it should remain listed. Where funds would come from if a five-year contract were approved was still unclear. After extended debate, agreement was reached to remove beryl from the list and to notify India that if a one-year renewal were not possible, then the contract should be allowed to expire in 1955. The Indian government accepted the one-year renewal, extending the contract on beryl to 30 September 1956, and followed this notification with a formal letter on the thirtieth.[25]

In this instance the government of India took what it could get, perhaps aware that given the number of other thorny issues then being negotiated with the United States and the gains from the thorium deal, it should not push the issue. It was, of course, this time confronted by a take-it-or-leave-it attitude, and India had no other ready buyers. Yet the repeated refusal of the AEC to come up with the necessary funds and its willingness, though reluctant, to abandon its efforts to restrict trade of beryl were symptomatic of the change that had occurred since General Leslie Groves had led the Manhattan Project. The era of the search for rare ores was not at an end, but the frenetic and costly phase was now of the past.

Such also was the message of negotiations with Brazil. The progress in uranium talks achieved in 1951 did not bring prompt movement the following year on a deal for monazite, thorium, and rare earth compounds. On 21 February 1952 the department of state informed the Vargas government that it was prepared to purchase the materials, subject to seven conditions. These had been previously worked out with the Brazilians and were considered in Washington as the basis for further negotiations on such matters as price. A Brazilian note of the same date indicated acceptance of these bases and expressed interest in selling the materials in question.

When an American negotiating team visited Brazil six months later to work out specific procurement contracts, a snag developed. According to William Freeman, special assistant to the commissioner of the U.S. Emergency Procurement Service, the problem was that the Brazilians could not reach agreement among themselves about the export of thorium. After reviewing production prospects, the Brazilian officials believed that their earlier estimate of production had been too optimistic. They asked for a one-year contract rather than the three-year term envisaged in the 21 February notes and asked that quantities not exceed the minimum amount considered in February. Brazilian producers, though they could not deny that making a three-year commitment would be unrealistic, nevertheless wanted the long contract in order to assure sale of thorium which otherwise could not be marketed. There were differences on price as well, with at least one producer willing to accept a lower price from the United States than was the government, again in order to assure at least some revenue from thorium sales.

The Americans returned home on 11 September without an agreement. Further negotiations eventually resolved price differences, and two procurement contracts were signed by the two governments on 23 December. These were for one year's duration, with a proviso stating that ninety days before their expiration consideration would be given as to whether further arrangements would be desirable.[26]

When the time came the following September to consider renewal, the department of state discovered that no U.S. agency was interested in buying more Brazilian rare earth compounds; only the AEC was interested in thorium. This news came as a bit of a shock to the Brazilians, who had need of the foreign exchange revenues that the sales provided. In Rio some officials referred to the February 1952 notes which spoke of a three-year term for the intended contracts. The Americans carefully pointed out, however, that the one-year term had been mutually agreed upon and particularly at the wish of the Brazilian government. Though influential Brazilian interests might be disappointed and cause undesirable repercussions on such matters as the "long-continuing uranium problem," the state department could "not attempt to arrange procurement of materials which are not now needed in order to forestall this."[27] There could hardly be a better indication that uranium was no longer something that had to be procured regardless of cost.

Dispute over the obligation of the United States to continue the thorium contracts blocked any meaningful negotiations. The situation was further clouded by the serious condition of the Brazilian economy and the disappointment of Vargas' regime that the United States was not providing greater loans than the substantial amounts gradually being contracted. Food shortages in certain regions of the South American country brought requests for wheat sales in 1954. But how were these to be financed? In the spring of that year Secretary of State John Foster Dulles informed his ambassador, James S. Kemper, that the department was thinking of bartering 100,000 to 150,000 tons of wheat for strategic materials. By the end of the month the Office of Defense Mobilization had authorized its funds to be used in such an exchange of wheat for monazite and rare earth derivatives. On 20 August agreement was reached in principle. The notes of 21 February 1952 and the debate whether they consituted a specific commitment themselves or merely the basis for future negotiation of specific contracts were set aside. Brazil was to receive 100,000 long tons of U.S. wheat, and negotiations were to be started for the possible purchase by the United States of an equivalent value of monazite and rare earth sodium sulphates. Four days later Vargas, aging and ill and confronted by a serious domestic political crisis, committed suicide.

The negotiations for purchase of monazite and sodium sulphates nevertheless began in September but were suspended two months later. Efforts would continue to be made to arrange for funding of Brazil's wheat purchase. For the purposes of this account, it is sufficient to note that earlier, when wheat supplies would have been a trump card in American bargaining for Indian materials, the card was not played. In the 1954 Brazilian negotiations, wheat was offered and purchase of unwanted monazite viewed as a diplomatic means to help the Brazilians acquire the wheat. The department of state had evidenced some consistency in not letting its interest or lack of interest in rare ores and their death-dealing components affect Washington's distribution of life-giving food.

Efforts for Additional Agreements

The dealings with India and Brazil were almost more trouble and cost than they were worth. Progress was spotty elsewhere as well. A sample of high-grade uranium led to the initiating of talks with the government of Argentina, but its ideas of prices "seemed unrealistically high."[28] In February 1951 the AEC asked the state department to discuss possible explorations for uranium with the governments of Bolivia, Chile, Colombia, Peru, and Mexico, in addition to Brazil, yet two years later not a single firm agreement had been reached for widespread exploration and sale of uranium, and only the representatives of Bolivia and Columbia appeared receptive.

The United Kingdom's agreement with Portugal, scheduled to terminate in 1957, restricted the size of the Portuguese uranium program and limited exports to 100 tons per year. The CDA countries had learned, however, of promising new areas for exploration and hoped to arrange a more aggressive program. The all-powerful Salazar held little interest, apparently willing to leave the ore in the ground until prices rose. The subject of uranium was not even broached with Spain in 1952, pending conclusion of military aid discussions; details of a working program for visiting geologists were to be planned in 1953.[29]

Canada offered better prospects. Negotiation of additional ore contracts had bogged down in 1949 in a conflict between the Canadians' desire for greater profits and the reluctance of the American Atomic Energy Commission director of raw materials, John Gustafson, to abandon the old cost-plus system or to allow greater discrepancies with the price paid to the Belgians. He was succeeded by Jesse Johnson on 1 January 1950. Johnson listened to the military's call for uranium at any price. Shipments from Africa and South America could be interdicted, not so those from Canada, which made the latter all the more worth cultivating. Rumors were also percolating of a vast uranium find in Athabasca. Accord was reached in March 1950 for a new contract for 8,000 tons of Canadian oxide to be purchased over the next eight years. A complicated price formula avoided conflict with the Belgian schedule. That is, the price of $2.75 per pound of oxide obtained from 1 percent grade ore stood, but the Americans required a minimum extraction of just 70 percent, effectively increasing the price per pound to $4.45. They also added a millage allowance of $7.25 per ton of raw ore. Though Canadian production through 1958 was pretty well committed by this contract, Canada was allowed to reserve some uranium for her own purposes.[30]

The new contract made possible exploration and development of the Athabasca reserves through the profits obtained by the government's Eldorado company. The Americans were after more than this. They wished to encourage private Canadian uranium developers as well. Low interest loans could serve as an enticement, but neither the Canadians nor the Americans liked that approach. Higher prices were needed and announced the beginning of March 1950, not without confusion between Washington and Ottawa on amount and timing. The proposal of American loans was forgotten; the Canadians maintained their policy that Eldorado would serve as the intermediary between the private companies and the United States, and the entrepreneurs set to with a will.

C. D. Howe nevertheless remained lukewarm to the sudden expansion of uranium production that the Americans were pushing in conjunction with their effort to produce a dominating number of weapons. His concern was for the development of the Eldorado holdings, not those of private businessmen. To peak his interest, Dean arranged a tour of the vast United States nuclear engineering facilities for Howe and other Canadians in August 1952. The experience gave Howe a new feeling for the effort, complexity, and expense the Americans were putting forward to assure the security of the West. He responded favorably, agreeing in September that Eldorado would sell uranium to the United States for the next ten years as long as the price was as good as anywhere else.[31]

News from France and Australia was also promising. Spring of 1952 saw the drafting of an agreement for a joint exploration and development program in French Morocco. The American Atomic Energy Commissioners were not happy that 20 percent of the uranium would be retained by the French but saw the deal as the best they could get. Relations with Australia were cordial, and the CDA struck agreements in June of 1952 for exploration and development of deposits at Radium Hill and in January 1953 for those at Rum Jungle. Here again there was a reserve clause: three years' worth of ore should remain in the ground to provide for eventual Australian needs. Disagreement in price only briefly slowed the negotiations. The Australians initially asked for $11.00 per ton, which the CDA refused to pay because of its implications for agreements with Belgium, Canada, and South Africa. Predictions were that Rum Jungle would produce 200 tons of oxide in 1954 and double that amount the following year.[32]

Of all the countries with which the CDA was dealing, South Africa offered the best hope of increased uranium shipments. Even before the November 1950 accord had formally been signed with South Africa, the Combined Development Agency was striving to augment its purchases.

The original agreement with South Africa called for recovery of approximately 1,200 tons of uranium oxide per year at six mines. American representatives now inquired about raising production to as high as 3,000 tons per year. This, it turned out, would create serious problems for the South Africans, and the Americans saw the need to increase prices and to offer other inducements. The South African mining industry, at CDA request, then conducted a thorough feasibility study which concluded that the goal of 3,000 short tons of uranium oxide was probably achievable by the close of 1956.

Negotiations for a "Supplemental Heads of Agreement and Addendum" proceeded forthwith between the CDA and the South African Energy Board. The price increase eventually accepted ranged from 7 to 18 schillings per pound of U_3O_8, or about $1.00 to $2.50. Under the new contract the cost of an average pound of oxide therefore rose from $9.10 to $10.50. The $14.00 (100 schillings) per pound ceiling price was retained. A new promise was that if the ceiling were reached, the mining company could ask to discontinue production, and the CDA could either accept that discontinuation, negotiate a new price, or require that the company continue mining on the basis of cost of production plus a profit margin of 10 schillings ($1.40) per pound. Because most of the mines had acquired uranium production facilities through loans, it was agreed that if production ceased, the Combined Development Agency would take over the loan payments. The CDA also agreed to purchase a greater amount of uranium than it had under the 1950 agreement. A final commitment involved CDA assistance in arranging a loan for the South African Electricity Supply Commission so that it might develop additional power supplies equivalent to those required by the uranium program. By the end of the summer of 1952 the Export-Import Bank had authorized a near $20 million loan to the Electricity Supply Commission for this purpose. Another $35 million had already been provided by the same source to build several extraction plants. The first of these, at West Rand Consolidated Mine in Krugersdorp, came on line in September 1952.[33]

The supplemental agreement was initialled in South Africa. Formal government approval was apparent as far as the British and South Africans were concerned, and the CPC quickly added its own, as consideration was hurried in order not to delay expansion of facilities already under construction. So it was that uranium did not pose any serious problem of agreement for the three countries. The situation was different for manganese and chrome, since South African shipments of these materials had fallen off sharply in 1951 from what they were in 1950. Some officials in the African state were in favor of conserving the Union's reserve of these items; there were those in both the United States and the United Kingdom who thought the Americans should withhold shipments of steel and equipment needed by South Africa until chrome and manganese exports were augmented. This idea was not favorably received in the state department, where it was recognized that invocation of retaliatory measures would make the U.S. "immeasurably the losers."[34] South Africa had not only chrome and manganese, but also asbestos. Moreover, the Americans' interest in uranium production was such that they would not likely consider measures which would negatively affect that program.

Concern for uranium, manganese, and chromium supplies helped further discussion in the state department as civil unrest grew in South Africa. There, rising racial differences added to those between English and Dutch settlers. The Malan government's vigorous application of *apartheid* drew widening criticism. Assistant Secretary of State for European Affairs George Perkins wrote Secretary of State Acheson in June 1952 that:

> Any serious disturbances in South Africa could have a direct bearing on our ability to get manganese, chrome, and uranium. It would have a damaging effect on Commonwealth relations and present great difficulties in our relations with South Africa. The future course of developments in South Africa is unpredictable. It is clear, however, that the situation is disturbing and that South Africa is heading for serious trouble.[35]

The perception was correct, the outcome still unclear three decades later.

Amid the efforts to obtain new or expanded arrangements the Combined Policy Committee did not overlook the obvious: renewal of the agreement with Belgium. Depleted though the Congo reserves might be, they could contribute to the AEC's import goal of 12,500 tons of oxide per year. Sengier was approached during an October 1952 visit to the States. To avoid annoying him and Robilliart, a planned military inspection of security at Shinkolobwe was canceled. The talks went well. In December Dean contacted Ryckmans, who soon responded that he and Prime Minister Van Zeeland were prepared to discuss renewal of the Tripartite Accord. Dean knew the Belgians would expect substantially more scientific information than they had received in the past. He warned Secretary of State Dulles that, while the AEC and Britain had previously assisted the Belgians "in a variety of ways to establish a small atomic energy project" by providing declassified information, further aid would have to come from classified areas.[36]

Formal talks began in April 1953 but snagged on scientific exchange. The small country desired help in building, not just a small research reactor, but a full-scale power plant; this the American Atomic Energy Commission and National Security Council were willing to grant. But promises meant nothing until the Atomic Energy Act was amended. Negotiations were suspended. At the end of the year Belgian leaders were upset that they were not consulted prior to Eisenhower's presentation of his Atoms for Peace plan to the United Nations. After all, some of the fissionable material the United States would be donating to the international pool originated in the Congo; once again, though Belgium was supposed to have an inside track to nuclear power, she seemed to be treated in a manner similar to other states which had provided nothing to the development program. Admiral Strauss, who became chairman of the AEC in July 1953, acknowledged the problem and urged that the first power reactor built abroad with United States assistance be located in Belgium.[37]

On August 30, 1954, President Eisenhower signed the long-awaited amendment to the Atomic Energy Act. Earlier, the NSC had approved a statement on cooperation with other nations in the peaceful uses of atomic energy which called for a program of aid in construction of reactors in selected nations, Belgium being the first; it noted that "the Belgian precedent should prove valuable background for formulating U.S. policy in respect to other nations."[38]

The diplomats set to work promptly. When Spaak came to the United States for the ninth session of the UN General Assembly, a major meeting of Belgian, British, and American representatives was held in Washington. There Strauss presented the foreign minister a paper on principles to be included in a "Memorandum of Cooperation between the United States and Belgium." The British representative submitted a similar informal note, and the two men explained that they desired separate Belgium-United States and Belgium-United Kingdom agreements, even if identical.

It had been understood for several months that the United States would provide technical aid in the construction of a Belgian power reactor. With that major issue already settled, discussion focused on how Belgium's future needs would be treated if halfway into the ten-year contract period it became apparent that her uranium was no longer needed for the strategic defense of the free world. John Hall of the AEC stated that Belgium's needs would automatically be met, in that the United States agreed to sell Belgium what she required in uranium metal and enriched material for ten years, even if the ores of Shinkolobwe were exhausted prior to the expiration of the contract. The willingness of the Americans to sell enriched uranium marked a sharp change from their past policy, but it did not fully meet the Belgian concern. Ryckmans explained that what was desired, especially in light of questions which might arise from the chambers, were reemption rights for uranium (if not needed for defense) to stockpile against the day the mine bottomed-out. He thought Belgium should either retain a portion of current production or permit that portion to be stockpiled in weapons form in the United States, subject to repurchase by Belgium. Spaak, for his part, thought there should be some provision that Belgium's obligation to supply uranium would halt in the event of a disarmament treaty.[39]

No resolution could be reached at the meeting. Afterward, Strauss asked his associates to draft a provision to the effect that "should the circumstances of world utilization of atomic energy be substantially altered during the life of the Agreement," the governments would "undertake to reexamine the basis of the arrangement."[40] At lunch Spaak indicated such wording would ease his parliamentary problem. It did not entirely meet Ryckmans' desire to begin a stockpile, so further adjustments were later made. The CDA already had agreements with Australia and France permitting them to hold some uranium in reserve; it would be difficult not to grant a similar right to the Belgians and of course embarrassing if it were to be denied and word of the other contracts leaked to Brussels. So it was that agreement was eventually reached that the United States would purchase 90 percent of the Congo uranium production for the first year following the expiration of the old contract in February 1956. From 1957 to 1960 the United States would purchase only 75 percent of the production. Belgium was assured access to American industrial nuclear information equal to that enjoyed by Great Britain and Canada.[41] Though not part of the Combined Development Agency, Belgium would be something of an "associate" of the club. No information is available regarding the last years of what was originally planned to be a ten-year accord. The percentage for the later years may have been left to be determined, or perhaps the accord was limited to only five years, as Ryckmans once implied in the 12 November gathering.

Belgium continued to receive considerable tax revenues, yet these were to decrease. By 1953 production had fallen to approximately 970 metric tons of uranium per year and would rise only slightly in 1956 and 1957. In 1958 production zoomed to 1,822 MTU; in 1959, the last year before the granting of Congolese independence, it was 1,784 MTU. Production then fell to 915 MTU in 1960 and to only 123 MTU in 1961 as civil war wracked Katanga and the Congo and the Shinkolobwe mines neared exhaustion. Productivity since then has not meritted recording in standard surveys.[42]

The Linkage of Preemption and Proliferation

The tale of the diplomacy of uranium and thorium acquisition in the early 1950s is essentially one of the unravelling of the concept of preemption and of the proliferation of nuclear knowledge and capabilities. The very effort to obtain a duopoly of uranium and thorium undermined the concomitant effort to limit proliferation of nuclear knowledge. The two were linked by the paucity of initially known resources in the United States and United Kingdom. The second had to be traded for the first; money alone would not do the trick. Promises were made; eventually processing plants were built, unwanted materials purchased, and the 100 percent control of each nation's uranium and thorium envisioned in 1944 and 1945 abandoned. By 1954 the state department was arguing that the way for the United States to maintain its leadership, tighten alliances, and influence neutrals such as India was not to withhold information but to take the lead in helping other nations build power reactors. Had Russia not made such rapid progress in her program, it is unlikely that the Americans would have been willing to share even as much as they did in the middle 1950s. Motivation to give information to other powers came, not from generosity, but from the arms race.

The technology was disseminated so slowly, however, that Britain and the United States reaped resentment as well as thanks. The stress of debates over sharing of information and material took its toll on the collaborative efforts of the Combined Development Agency as well; it is significant that the planned renewal of the 1944 Tripartite Accord took the form of two bilateral treaties. Much had changed in a ten-year period. As President Eisenhower proclaimed in his Atoms for Peace address, a new era in the international handling of nuclear technology and materials was beginning. Though he attempted to move his country into it in a positive, forward manner, in actuality the United States moved hesitantly and reluctantly. It backed across the frontier more than assertively broaching it, nudged on by Britain, Belgium, even India, by the American thirst for uranium and desire to strain Russian reserves, and by the progress of the Soviet Union.

10 · Preemption and Monopoly in Retrospect

An account of the British and American hunt for uranium begins and ends most appropriately with discussion of accords with Belgium. The Belgian Congo connection was at the heart of the search which led negotiators to all continents except Antarctica. The issues of information sharing, proliferation of knowledge, security of the West and of the Congo mines, reservation of ores for future national use, distrust of Germany and Russia, and price, as well as the roles of a multi-national corporation and the ore itself, were involved in these central agreements. If over the decades since the outbreak of the second world war Belgian uranium represented only a low percentage of American and British purchases of that commodity, it was the Congo's 17,000 to 18,000 short tons of uranium oxide produced in the 1940s and early 1950s which were most significant for the arms race. Canadian uranium was in the first atomic pile at Chicago. But Belgian ores fueled subsequent key experimental work and the bombs that hastened the end of World War II. Primarily Belgian uranium provided the nuclear deterrent which was repeatedly used by the West as a counter-threat to Stalin's divisions. The wisdom of reliance on such a threat may be debated. Yet it is clear that in the crucial postwar years the success of the Western diplomats and military statesmen in obtaining the single best source of rich uranium deposits was central to the entire defensive posture of the West. One need only to imagine the converse of the situation—the Western powers lacking the raw materials to produce their weapon and the concomitant implications for military manpower policies and the several crises relating to Berlin— for the significance of this ore to be acknowledged. The symbol chosen by the Belgians for the World's Fair hosted by them in Brussels in 1958 was well justified: the spectacular "atomium" rising in silvery geometric form over the gathering of exhibits from all sectors of the globe.

The Tripartite Agreement of 1944 encouraged Groves and his successors to conceive of the possibility of gaining control of the world's key supplies of uranium and thorium. Yet it is difficult in retrospect to understand how preemption could be expected to work. Though

Byrnes could say to the scientist Leo Szilard in 1945, "General Groves tells me there is no uranium in Russia," such a view, given the lack of knowledge of the Soviet Union's resources, was optimistic at best.[1] Years later Lilienthal would state that covert agents in Russia reported no uranium.[2] Perhaps they referred only to the sort of rich deposits which for a time were believed to be the sole type practical for successful bomb production. The Soviets' security precautions were extensive, and the agents may have been misled. Szilard warned that the Russians surely had low-grade ores which could be used. So too did James Franck, an emigré German scientist who served as head of a special Committee on Social and Political Implications of Nuclear Weapons formed at the time of the debate over whether the bomb should, in fact, be used. His report read in part:

> It may be asked whether we cannot prevent the development of military nucleonics in other countries by a monopoly on the raw materials of nuclear power. The answer is that even though the largest now known deposits of uranium ores are under the control of powers which belong to the "western" group . . . , the old deposits in Czechoslovakia are outside their sphere. Russia is known to be mining radium on its own territory; and even if we do not know the size of the deposits discovered so far in the USSR, the probability that no large reserves of uranium will be found in a country which covers 1/5 of the land area of the earth (and whose sphere of influence takes in additional territory), is too small to serve as a basis of security. *Thus, we cannot hope to avoid a nuclear armament race either by keeping secret from the competing nations the basic scientific facts of nuclear power or by cornering the raw materials required for such a race.*[3]

Conant and Bush for their part advised Stimson as early as the fall of 1944 that "control of supply" could not be counted on over a period of a decade to prevent other countries from making the same secret developments as had the United States. Creation of a hydrogen bomb would make supply control even less effective, as only a small amount of uranium or thorium would be needed as a detonator.[4] Gordon Dean would write in 1953 that Soviet control of the Erzegebirge region of Saxony and Czechoslovakia, including the Joachimstal mine "alone could support a sizeable atomic energy program."[5] Fear of German control of the Czech mines had stimulated Albert Einstein's original letter to President Roosevelt about bomb development. Passage of these mines to Soviet control surely did not diminish their capacity as a source of ores. Moreover, the Russians held many low-grade ore deposits which they could develop just as the Americans had developed theirs.

Just how severe the materials situation was for the U.S.S.R. is not known in the West. There are indications that the growth of the Soviet Union's bomb production, if not its development program, was slowed by the CDT's monopolization of non-Communist world uranium supplies. A Soviet book published in 1955 (A. A. Santalov's *The Imperialist Struggle for Raw Material Resources*) complained of the U.S. monopoly of uranium supplies, and *Pravda* raised even stronger objections.[6] In fact, the backers of the preemption efforts did not aspire to forestall German and Soviet bomb development indefinitely, but rather to delay it; once a nuclear weapon was achieved by the Russians, they hoped so to starve the Soviet production rate, especially in comparison to that of the United States, that the West would hold a significant preponderance. In this they appear to have succeeded for at least a few years. Correctly or not, American and apparently also British officials believed that Soviet supplies were so limited that the West could establish what was hoped to be a long-term *pax atomica*. Ironically, this may eventually have come about, not because of the preponderance of nuclear weapons on any one side, but by a balance of these and their delivery systems.

In his memoirs James F. Byrnes tells of a spring 1945 meeting with the Interim Committee which included several scientists as well as military figures. At that time the consensus was that the Soviets would have the secret of the bomb in two or three years and require six to seven years to produce the bomb. "No one seemed too alarmed at the prospect because it appeared that in seven years we should be far ahead of the Soviets in this field. . . ."[7]

Some writers have suggested that top American officials believed an early wartime estimate that it would take the Russians twenty or more years to catch up with the West; because of this belief American diplomacy at the war's end was defectively tailored.[8] Especially dangerous, says Gregg Herken, was faith in the myth of an atomic secret which, if there were any at all, was not so much technology as the monopolization of uranium supplies. The failure to hear the warnings of the scientists regarding the inevitable and likely prompt development of rival atomic weapons is attributed to Groves' arrogant attitude toward scientists and the success of his efforts to keep their opinions at some remove from top policy makers. But U.S. officials did take initiatives for international control and soon recognized that they were operating in only a limited and shrinking grace period. Moreover, at the time their British partners were even more opposed than they to any sharing of information with third parties. Churchill strongly advocated secrecy, especially as far as the French and Russians were concerned. Belief in the success of the CDT's efforts at control of uranium supplies may indeed have caused both unfortunate delay in consideration of positive atomic policy and overconfidence in the capacity of Western technology to shape the postwar world. The death of President Roosevelt and the multitude of problems involved in ending the war also played their role in keeping atomic energy policy from being given priority of place.

The concept of preemption, the practicality of which Chancellor of the Exchequer Sir John Anderson never was convinced and which also was occasionally questioned on the United States side, first appeared in regard to thorium. The intent for uranium was to buy as much as possible for conversion to warheads as fast as possible. The goal was to obtain weapons fuel for the CPC countries rather than to deprive a rival.

This posture did not last long, whatever the public statements. The effort for a 99-year lease of Congolese uranium surely was a preemptive move. But it was the discovery in 1945 of the resources hidden in Swedish kölm that pushed the Americans and British forthrightly into preemption of uranium as well as thorium. The active concern over kölm reflected recognition both of the Soviet atomic effort and of the possibility that ores of a much lower grade than those of the Congo could be used at a price. Thus preemption was not long conceived—if it ever was—as a means of *preventing* the Soviets from gaining the bomb. It was, however, seen as a way of delaying mass production and greatly raising the cost of such production to the Soviet economy. As such, the materials search became an early and aggressive feature of the Cold War. Nor did it disappear quickly, for Eisenhower's Atoms for Peace proposal had the effect of either stripping the Russians of some of their still limited stocks or forcing them publicly to refuse participation in an international sharing of nuclear materials, thus damaging their image.

The effort to establish a uranium and thorium duopoly was, of course, a gamble. The odds were high against it, yet the possible long-range power benefits seemed worthwhile to people like Groves and even Makins. Groves, never known to be sensitive to the feelings of others, gave only slight consideration to the impact success or failure would have on other power rivals. Success would promote a sword-of-Damocles tyranny bound to build resentment and resistance; failure would demonstrate to the rival, first, the need to press similar arms development as quickly as possible and, second, the apparent treachery and greed of the opponent. Emerging states would see their exclusion from the "club" as a further demonstration of colonial capitalist power domination and view the sale of their own ores as opportunity to gain status and respect as well as dollars, information, and goods.

The decision to strive for a monopoly of the rare ores was reached by the CPC very early, when both the necessity and the scarcity of rich uranium deposits were generally accepted beliefs. The policy was put forward by Groves and adopted without much debate. Once on line, it was pursued energetically by Groves while its implications were not extensively considered. By the time the state department was informed, execution was well under way. Throughout the years in question, the state department never had more than one specialist on atomic affairs at a time. As an assistant to the assistant secretary of state, Gullion combined these duties with others. The efforts of his successor, Gordon Arneson, were more exclusively devoted to atomic energy affairs, and his position eventually was elevated to that of special assistant to the secretary. For the most part Arneson's views were unchallenged, as his superiors were occupied with other matters and there were no other specialists on the same topic within the department to question his posture. His opinions on uranium matters were close to those of Groves, if expressed more diplomatically. This was natural, for Arneson was originally tapped by Groves and shifted to the state department directly from Groves' military stable.

The implications of the U.S. war department's initial total control of negotiations should not be overlooked. Roosevelt was informed and individual state department members involved in particular negotiations. Yet until the passage of the Atomic Energy Act of 1946 the traditional civilian reviewers of national foreign policy, the top officials of the state department and the U.S. senate, were either prevented from a fully informed discussion and evaluation of American policy on uranium procurement or kept in the dark entirely. Dean Acheson and Arthur Vandenberg had their differences. However, their outrage when they learned of the various agreements stemmed from a common concern: in a democracy basic commitments regarding foreign policy and the lives, fortunes, and security of the population, especially in peacetime, should be under the purview of the duly constituted civilian authorities; they should not be controlled in secret by the military responsible to just the chief executive. The matter was one of principle. The raising of it did not necessarily impugn the ability, integrity, or principles of the military (after all, who could challenge George C. Marshall on these counts; of course the crucial Quebec Accord was made by the president). Nor did it necessarily involve alteration of policy; Vandenberg, Hickenlooper, and McMahon proved equally as ardent hoarders of uranium and information as Groves.

The practice of military figures negotiating accords, with and without the involvement of individual foreign policy officials, is not new and has frequently been justified by existing circumstances. Was this so in this case? Surely during the war years, but not so clearly thereafter. The vision Groves and Stimson held was one of safety—but concomitantly of world military supremacy through control over two commodities, uranium and thorium. In short, a policy with sweeping implications for international relations and world peace was adopted by military officials—it is true, in conjunction with a few members of the executive but without normal review. Though congress was not aware at the time it approved the McMahon Act of the commitments which had been made, the act did serve to reestablish the concept of civilian control of policy. The price paid was resentment within the military establishment and hindrance of good diplomatic relations on atomic matters between the United States and Great Britain and Belgium.

The story of the diplomatic quest for uranium supply demonstrates well the intermingling of foreign and military policy with domestic politics, private company economics, and scientific affairs. The point requires no belaboring; it is neatly represented in Gullion's somewhat panicky note to Marshall at a time when Spaak's political position was precarious: "a change of government in Belgium would jeopardize our whole economic ore procurement program."[9]

In retrospect, the effort and the expenditures associated with the gathering of uranium and thorium appear excessive, but in the context of the times they were not. The American joint chiefs of staff wrote in 1949, while recommending an acceleration of the atomic energy program, that they were

of the opinion that the gain from the military standpoint of the proposed accelerated program over that which can be obtained from a continuation at the present level appears very significant in terms of lower unit cost of weapons; probable shortening of a war; increased military effectiveness; decreased logistical and manpower requirements for the prosecution of certain tasks in war; and increased flexibility in the conduct of the war, which is extremely important in view of the many imponderables now facing our planners. Furthermore, when the USSR attains a stockpile of atomic weapons, overwhelming superiority of our own stockpile and production rate will be necessary if our atomic weapon posture is to continue to act as a deterrent to war.[10]

The notion of the single weapon which could win a war against Germany or Japan or of an unmatchable and overwhelming stockpile of such weapons that could stifle Soviet ambitions was a will-o'-the-wisp. Yet the dream inspired important efforts and achievements from the day Roosevelt decided to act upon Einstein's advice. If the dream were potent enough to produce the bomb, why should it not be potent enough to inspire visions of hegemony? Wise heads said that this would not be possible in terms of theory, yet Groves appeared to have stumbled upon a pragmatic fact—the scarcity of uranium—that could nevertheless make it come true in reality. Industry, speed, an open purse, and above all, as Groves put it, "secret diplomatic agreement secretly arrived at" seemed the keys to the goal of supremacy.[11]

The effort to obtain a monopoly of uranium now appears as a search for security at best, megalomaniacal at worst. Surely the American efforts to achieve international control of the atom while secretly striving for monopoly of the key fuel appeared as capitalist imperialism and duplicity to the Soviet intelligence agencies. Whatever Maclean reported to the Russians must have nurtured their distrust, chilled relations, and fostered the Cold War. Maclean's treason aside, the activity of geologists and ore ships could not be totally concealed. Yet if the Anglo-American two-track approach to use of the atom appears as contradictory from one viewpoint, it nevertheless made some sense. The high road was international control. To achieve this, should the British and Americans have automatically discarded their lead in control of uranium resources? This would have been foolhardy in an era of *realpolitik*.

The entire issue of control of atomic power was further complicated by the atmosphere of hysteria in the West regarding Soviet ambitions and the mystique in which atomic power was enshrined in the public mind. United States relations with Belgium and Britain were especially intricate because, as Under Secretary of State Robert Lovett told the Joint Congressional Committee, "uranium has come to have such a symbolic value, bound up with national prestige."[12]

The problems associated with the sharing of information and raw materials caused serious difficulties in the Western alliance. Even as late as the Bermuda Conference of 1953, Churchill was reported to be "still very much rankled" by what he considered the Americans' breach of promise; Makins warned that "Churchill continues to brood about all this and is now in quite a bitter frame of mind."[13] The British and Americans did manage to collaborate on many fronts. Yet the difficulties experienced on atomic matters still can be found in the tones of annoyance in British accounts of the issue. The differences which developed were all the sharper because they were bolstered by a sense of justified righteousness. The work of British scientists did forward development of the bomb, and the Americans' unwillingness to share their later findings seemed a purposeful effort to push England to the status of a second-level power. Were it not for British assistance in the search for uranium, there is no question that the Americans would have been even more obstinate and ungenerous in sharing information. The quest for materials therefore mitigated what was to all observers a nasty diplomatic situation. The betrayals of spies of course only exacerbated, indeed nearly killed, the information exchange issue. The leak of information to the Russians also had the concomitant effect of making a monopoly of uranium supplies seem all the more necessary since the cherished scientific secrets presumably had been revealed.

But was there actually a race for uranium ore supplies? The some 3,500 tons of uranium compounds in Belgian stockpiles captured by the Nazis were several times greater than the supplies the Manhattan District had available to produce the first two American bombs. Uranium was also being produced at the former Czechoslovakian mines at Joachimstal. German industry was able to produce uranium metal at a maximum rate of close to a ton a month by the beginning of 1942. This was several months before such production was achieved in the United States. In short, the Germans had sufficient uranium to produce a bomb had their scientific efforts been successful. There was no German drive to acquire uranium matching that of the allies, in part because of Adolf Hitler's limited interest in uranium and in part because the German supply seemed sufficient. The competition with Germany was not one for uranium (heavy water was a different matter) but of science.[14] Despite good espionage work, Groves could not be entirely sure of the level of German atomic technology.

Soviet competition for the same ore supplies that attracted the United States and Great Britain appears to have been minimal. A Russian approach was made to Belgium which was quickly rebuffed, and there are no available reports of encounters with Russian survey teams or the like. The Soviets did exploit the resources of Saxony and Czechoslovakia and the shales of Estonia, but they did not compete with the United Kingdom and the United States in the world market. They would have had great difficulty in doing so, in any case, given their shortage of hard currencies and reluctance to sell gold. The extent of the presence of uranium, for example in Canada and the American West, and the practicality of mining low-grade ores did not become clear to the British and Americans until the mid-1950s; until that time they may justifiably have thought U.S.S.R. nuclear efforts were or would be hindered by supply problems.

If lulled, albeit only partially, by belief that their success in cornering the free world's uranium supplies precluded a weapons race, the Western allies were soon enough disabused of this view. At best it had been only a fleeting hope linked with the original primitive technologies that indicated great supplies of rich uranium ore were necessary to produce a bomb that would work. The very success of the United States in producing a bomb, heavy as it was, which a plane could carry over Hiroshima signalled to the world that such a weapon did not require inordinate amounts of fissionable fuel.

Mass production of atomic weapons did not occur rapidly. Lilienthal reports his own shock and that of President Truman when he learned, upon becoming chairman of the American AEC in 1947, that the United States had no complete bombs on hand and a stockpile of no more than 13 unassembled bombs. Plutonium production from the Hanford, Washington, reactors had been about halved since the war's end, dropping from 165 kilograms per year to 80 kilograms—only enough fuel to make about 13 bombs per year. Orders were given to increase plutonium output. Research on the creation of warheads utilizing both plutonium and U-235 to augment explosive yield and improve efficient use of raw materials accelerated. The development of the Redox process of chemical separation, which allowed recycling of uranium, was similarly pushed. The new efficient Mark IV implosion bomb tested in 1948 promised a 63 percent growth in stockpile numbers and a 75 percent increase in explosive yield over the previous model.

In October 1947 the joint chiefs of staff informed Lilienthal that growing security dangers required more weapons—this was the year of the Communist coup in Czechoslovakia, growing tensions over Berlin, and perceptible decline in the effectiveness of nationalist Chinese troops battling those of Mao Tse-tung. The joint chiefs called for a stockpile of 400 bombs of 20-kiloton yield by January 1953. The AEC saw this goal as possible and one it could meet even as early as 1951. As ore purchases increased and new facilities came on line, bomb production accelerated. By the close of 1949, 300 had already been stored. Authoritative estimates set the figure for the summer of 1953 at 1,000 and at 18,000 by the close of the decade. The expansion of material supplies and production facilities so long desired and considered difficult of attainment had, in fact, been achieved by the mid-1950s.[15]

The production rate reached by the first months of 1949, together with what the Americans and British considered a substantial technology lead over the Russians and the assumed Soviet shortage of uranium, especially of high-yield ores, led a number of Western officials to anticipate that their countries would hold an overwhelming nuclear hegemony for some time to come. News of the Soviet atomic explosion of late summer 1949 did not gravely threaten that assessment, as estimates were quickly made to the effect that, following the success of that test, the Soviets could produce only 20 bombs at a maximum in the following year and 135 at the most by mid-1953. Word of the treason of Klaus Fuchs, however, was more unsettling. Perhaps the Soviets had begun nuclear research earlier and with more success than the Americans had thought. Generals Herbert Loper and Keith Nichols warned the Military Liaison Committee in February 1950 that, if this were the case, thanks to Fuchs' assistance, then Soviet bomb production might indeed be close to that of the United States and that the Russians might be well along in developing a hydrogen bomb.[16]

To what extent the news of the Fuchs betrayal influenced the American decision to build an H-bomb is still debated. It is clear that estimates of Soviet weapon production were revised to bring the totals forward by about a year, as new guesses suggested that the Russians would have approximately 120 atomic weapons by mid-1952.[17] In American eyes, there definitely was a race on now. It was not for uranium but rather was one of design and production capacities. In May of 1952 the uranium supply had in fact so improved that Secretary of Defense Lovett was no longer arguing that ore procurement should be stepped up to meet weapons needs, but that "atomic plant capacity [should be built] to the level justified by the uranium prospects."[18] He was also arguing, in regard to uranium, that "it would appear common prudence to bring into the continental confines of the United States any and all strategic materials against the day that they might be needed."[19] So intense had the quest for uranium become that for Lovett production of weapons no longer commanded the search for new supplies; instead, the discovery of new resources seemed to justify the production of more weapons. The National Security Council, however, recommended that the AEC set its production goals only after the defense department had first stated its numerical requirements for atomic weapons.[20]

What if there had been open-market competition for uranium? What if, as Prime Minister Nehru evidently did in the case of thorium nitrate, leaders of the producer nations had put their materials up for the highest bidder? The price might have risen, but it does not seem likely that the end result would have been much different. Cost was never the determining factor. The French might have gained better access to uranium, rather than being forced to develop their own resources, which eventually proved to be considerable; and they might have exploded their test bomb earlier. Surely their annoyance at being shut out of the "atomic club" and their later determination to build their own nuclear force, independent of any NATO force influenced or controlled by the Americans, have some relationship to the Anglo-American reluctance to share information or materials.

Belgium did raise her demands after the war, but in a different manner than did Nehru or the Brazilians. Van Zeeland's negotiations, though long and painful, were essentially not to gain increased revenues for his country but simply to gain what was due it and early agreed upon in Article 9(a) of the 1944 accord. The United States and the United Kingdom were fortunate to be able to deal with individuals such as Spaak and Sengier, Gutt and Van Zeeland. Their basic commitment to Western security was prompt and clear; if the Belgians negotiated sharply, they also negotiated with a sense of responsibility. As Lilienthal told American members of the Combined Policy Committee, Spaak was a "tower of strength" for the West in this matter as in others.[21]

There can be little doubt that the statesmen of Belgium were hurt and angered that their key role in supplying uranium to the Combined Development Trust did not put Belgium in the forefront of the European atomic energy field, enabling her to run her factories on cheap electricity from atomic power plants and to export power to neighboring states. The vision of Belgium as a nuclear research leader and power supplier for Western Europe was a glorious one. So little was known of the technology of harnessing the atom and hopes regarding its beneficence so great, so ballyhooed, that all sorts of improper expectations were aroused. The failure of these hopes was not the fault of Gutt, Spaak, Anderson, Stimson, or Groves. The problem was one, first, of technology; second, of the Cold War; and, third, of the great concern which existed in the United States regarding control of this symbol and instrument of power.

As British author John Simpson has recently pointed out,

> The military aspects of nuclear energy . . . present the democratic state with an acute dilemma: how to reconcile the demands of those citizens anxious to obtain proof that the powers of the state have not been abused, with the genuine and sincere concerns of those guarding its secrets that such information could assist other states to acquire nuclear weapons and detract from national security. There also exists an additional dilemma for the state: how to reconcile the pressures to exploit nuclear energy for non-military purposes, especially in the context of an unregulated, free-market economy, with the requirement to prevent both information and materials being converted to military use either by non-state groups or non-nuclear weapon countries.[22]

Spaak, Van Zeeland, and other Belgians believed that more progress was being made on industrial uses of the atom than was being shared with them. The record of industrial development now shows that they were not being deceived. Industrial applications were slow to come in the United States, and every step of progress revealed additional problems. Under the guidance and persistent prodding of Admiral Hymen Rickover, the U.S. Navy did develop a nuclear reactor suitable for powering a submarine. This could be adapted to civilian purposes and was the sort of choice information the Belgians might well have thought they should receive in return for their earlier contributions. But the very utility of the reactor as a submarine power plant made it highly classified military information.

During the early years of nuclear development, any distinction between military and commercial usages of uranium was difficult to draw. Radioisotopes for medical research were an exception, and it was natural that Sengier and Groves relied upon gifts of these to alleviate criticism. Distinctions later to be made betweeen enrichment levels of uranium, between the higher level seen necessary for military usage and the lower, more easily reached, levels for civilian commercial usage, had not yet been recognized. One of the key arguments that McMahon and others would raise for the production of additional atomic bombs was that the rare uranium would be satisfactorily stored in the form of bombs and that, should later the bombs prove unnecessary, the uranium could be removed and without further processing be used for commercial purposes.[23] Thus there was some hypocrisy involved in American assertions in the 1940s to Belgian negotiators that the United States was not using Belgian uranium for domestic commercial and industrial purposes. It was not so being used at the time; but several officials thought it would be soon enough, and without alteration.

Be that as it may, the Belgians fared well in their dealings with Groves and the Combined Development Trust. Much of the credit for this goes to Sengier, who knew how to deal with the Americans and yet did not neglect the interests of his country and his company. A good initial price was obtained for the ore, and substantial capital improvements in the mine were financed by the CDT. Moreover, the premium added to the original price of the uranium to compensate for Sengier's loss of the Canadian radium market was quickly amalgamated into the permanent price structure. It no longer was viewed as a separate premium and did provide additional revenues to the Union Minière and, by way of taxes, to the Belgian government for more than a decade.

A more touchy question is whether the Congo received its fair share of the uranium profits. This ostensibly was not a concern of the CDT; its haste to exhaust the Shinkolobwe lodes suggests a desire to avoid any long-run risk involved in dealing for the ores with a possibly restive and independently minded colony. Surely the Belgian government could argue that it was pumping enough funds into Congo welfare and social services that the diversion after 1951 of uranium tax revenues to the atomic energy program in the mother country, the only path mutually agreeable to both the United States and Belgium to compensate for the Americans' non-fulfillment of Article 9(a) of the 1944 agreement, was merely a trade-off and not of significance. Just how the tax revenues were, in fact, allocated in the years following the 1951 agreement is not a matter of inquiry here and would require skillful reading of undoubtedly complex bookkeeping records.

Whether the uranium tax revenues in their entirety should have been devoted to the Congo, in addition to the other monies directed there by the mother country, is a different question, bound to be answered differently from a variety of points of view. To this writer, though no doubt not to the Belgian negotiators of 1951, the argument that expenditure of the tax revenues on atomic research plants in Belgium was justified in that it enhanced the value of the Congo ores appears pure subterfuge. As the British negotiator brought out at the time, whatever enhancement was going to take place had already been achieved by virtue of the American and British discoveries and plants. Moreover, all officials involved, and most especially Sengier, must have known that the Shinkolobwe ores would soon be exhausted. It should be remembered, however, that the Belgians were forced to this argument by the adamant American refusal to admit or acknowledge any form of payment which might suggest American acceptance of partial responsibility for the social and economic welfare of the Congo. This policy was well founded in traditional American distrust of colonialism, questions regarding the treatment received by the colony at Belgian hands in the past, and budgetary pressures for social reforms in the United States in excess of the funding available. Yet the U.S. refusal to funnel monies to colonies prior to their independence, compared with the generosity of finances made available afterwards, surely must have received wry notice in government bureaus in Brussels.

The Belgians were successful in their first goal of bolstering the defense of the West and in their second goal of gaining good financial compensation. They were less successful in gaining their third goal of research information which would give them a special commercial position in Europe. The low country did get a power reactor, but not significantly in advance of other nations. The Belgians failed markedly in a fourth aim, which was to obtain United States and United Kingdom aid for social reforms in the Congo. This last was never articulated as a clear goal to the allied negotiators. Yet the reaction of the American military leaders and the state department shows that the United States fully recognized a Belgian wish to entice American and British commitment to assisting the social and economic welfare of the Congo colony. The Americans understood that some Belgian officials also would have liked the two powers to accept a major responsibility for defense of the Congo, or at least for the defense of the province of Katanga, where the Shinkolobwe mine was located. The Americans did not bite, and though some military equipment was provided to the Congo, it was funnelled for the most part through Belgium, thus avoiding a specific major commitment to the defense of the Congo. Hopes, expectations, and perhaps even some plans for such defense may have existed in the minds of some individuals in Brussels and Katanga. The disappointment in those locales in 1960-1962, when President Eisenhower and then President John F. Kennedy refused to support efforts for the secession of Katanga from the newly independent state of Zaire, were therefore all the more bitter.

The patience of Spaak and of the British negotiators such as Roger Makins was badly needed in view of the domestic disarray in which the Americans found themselves in regard to protection of the atom. As mentioned earlier, the necessity of securing adequate supplies of ores was an issue which led men such as Hickenlooper, Louis Johnson, and LeBaron to moderate their attitudes on information sharing. Without the need for supplies, communication might have deteriorated even further to the point that future improved cooperation on atomic matters would have been even slower coming than it was. The state department, under frequent domestic attack in these years and soon to experience the full force of the hysteria inspired by Senator Joseph McCarthy, held amazingly firm. Acheson and Kennan prevented the European Recovery Plan from being linked to the uranium search effort and stymied efforts to starve other countries into agreement. Cooperation won by denying food supplies unless uranium were sold only to the CDT would surely have been grudging, bitter, and tenuous. Lilienthal in time had to withdraw from the battle as chairman of the AEC. As far as achieving an improved arrangement for information sharing, his successor, Gordon Dean, proved to be a most able negotiator.

The British were surprisingly slow to recognize the impact of their security lapses upon the diplomatic situation, just as they were slow to acknowledge their slippage as a world power. Indeed, their desire to obtain knowledge of American scientific developments was a mark of their effort to retain the great-power status of which now the bomb, more than their depleted fleet, was the accepted measure. Their view was longer and more sophisticated than the Americans'; their determination to work things out despite rebuffs was greater, again a reflection of experience as well as of need. Anderson was right about the difficulty and questionable utility of controlling thorium supplies, but the British went along with the American effort. Their desire to cooperate more with the Scandinavian countries was also probably correct. The British, however, gained far less directly from their participation in the uranium search than they might ultimately have expected. They did win much indirectly, for it was the Americans who gave much stimulus and funding to explorations in South Africa and Australia that would eventually provide key supplies to the British program. And they too, even as the Belgians, benefitted from the atomic shield provided by the United States in the period of severe British weakness following the close of the world war.

American diplomacy was brash; it was headstrong; it was idealistic and perhaps naive in searching for goals that were in reality unobtainable. Achievement of a monopoly of the world's uranium and thus of atomic power was simply not possible. But U.S. policy was not as selfish as might appear at first glance. The terms offered for uranium in each case were generous and usually extended in a far more patient and persistent manner than would have been expected, given the pressures of both hot and cold wars and mounting domestic debate over nuclear defense. Great effort, thought, and expenditure were extended on behalf of defense, not just of the United States and United Kingdom, but of all the Western countries. Industrial applications were unexpectedly slow in coming. Nevertheless considerable information was shared, particularly in the areas of medical applications and eventually low-power reactors, more than is sometimes recognized under the chorus of complaints that even more information was not made available.

The dynamism, vision, and heavy-handed behavior of General Leslie Groves will remain controversial; it is no doubt well that men such as Acheson and Eisenhower emerged to balance him. Sengier and Spaak played key roles, as did Anderson and Makins. Yet the final impression is not of individuals but of an overwhelming effort, so vast in its worldwide scope and expenditures, so vital in its implications for international relations and for the future development of mankind's technical abilities, that one is almost taken aback by the breadth of its ambition and the extent of its success.

NOTES

CHAPTER 1: DISCOVERING THE NEED

1. Leo Szilard, "Reminiscences," ed. by Gertrude W. Szilard and Kathleen R. Winsor in *The Intellectual Migration: Europe and America, 1930-1960*, ed. by Donald Fleming and Bernard Bailyn (Cambridge, Mass: Harvard Univ. Press, 1969), pp. 111-13.

3. June H. Taylor and Michael D. Yokell, *Yellowcake: The International Uranium Cartel* (New York: Pergamon, 1979), p. 23.

4. As quoted in Leslie R. Groves, *Now It Can Be Told* (New York: Harper, 1962), p. 33. Note of meeting of 8 May 1944, United States National Archives, Modern Military Records Division, Harrison-Bundy papers (hereafter cited as MMRD, H-B), 54.

5. David Irving, *The German Atomic Bomb: The History of Nuclear Research in Nazi Germany* (New York: Simon and Schuster, 1967), pp. 70-71.

6. Groves, *Now*, p. 180.

7. Robert Bothwell, *Eldorado: Canada's National Uranium Company* (Toronto: Univ. of Toronto Press, 1984), pp. 73, 96, 110. This history of the Eldorado company significantly supplements the official Canadian account of nuclear development in that country: Wilfred Eggleston, *Canada's Nuclear Story* (Toronto: Clarke, Irwin, 1965). Margaret Gowing, *Britain and Atomic Energy 1939-1945*, United Kingdom Atomic Energy Authority (New York: St. Martin's, 1964) and Margaret Gowing assisted by Lorna Arnold, *Independence and Deterrence: Britain and Atomic Energy, 1945-1952*, 2 vols., United Kingdom Atomic Energy Authority (New York: St. Martin's, 1974) constitute the official British histories. Richard G. Hewlett and Oscar E. Anderson, *The New World, 1939/1946*, and Richard G. Hewlett and Francis Duncan, *Atomic Shield, 1948/1952*, comprise the two volumes of *A History of the United States Atomic Energy Commission* (University Park, PA: Pennsylvania State Univ. Press, 1962-1969).

8. Hewlett and Anderson, *New World*, p. 86.

9. Ibid., and Groves, *Now*, p. 36. The controls did not take effect quickly enough to prevent Soviet purchase from the United States and Canada of small amounts of uranium oxide, uranium nitrate, and impure uranium metal. House of Representatives Committee on Un-American Activities, 81st Cong., *Hearings Regarding Shipment of Atomic Material to the Soviet Union during World War II* (Washington: U.S. Government Printing Office, 1950).

10. The principal parts of an atom are the proton, electron, and neutron. The element uranium has 912 protons and 92 electrons, but its atoms may have differing numbers of neutrons. Most frequently the uranium atom

holds 92 protons and 146 neutrons; this is called the U-238 isotope. Occasionally, less than one percent of the time, the atom will contain only 143 neutrons; this figure added to that of 92 protons gives another isotope its identity as U-235. Of the some 80 pounds of uranium produced from a short ton of 4 percent ore, only about .7 percent is in the form of the U-235 isotope. For a complexity of reasons, this isotope is less stable than other isotopes of uranium and is fissionable. The critical mass of U-235 necessary to sustain a nuclear reaction was one of the chief secrets of the scientific war. In any case, a great many tons of high grade ore were needed in the 1940s to produce a significant number of pounds of U-235.

Obtaining U-235 from uranium ore is not easy, and in the process the yield per ton of ore is further reduced. After milling, the uranium oxide concentrates in the form of yellowcake are converted to the heavy poisonous gas uranium hexaflouride at high temperatures. This gas is then processed through a cascade of filters at a gas diffusion plant to separate the U-235 atoms. The process is not an ideally complete one, with the result that the tails, or U-235 left in the waste produced by the gas diffusion plants, typically was .2 to .3 percent with the technology available as late as the 1970s. Further filtering can reduce this remainder and improve the yield, but only at great energy and construction costs. No doubt during the war years and immediately thereafter the efficiency rate was low, and the need for rich ores which would give significant yields was correspondingly high.

A possibility of easing the supply situation in the distant future was revealed by discovery in 1940 that fission of U-235 atoms could lead to the absorption of a neutron by a U-238 atom. The resulting U-239 promptly emits particles, becoming a new element with 93 protons and 93 electrons. Called neptunium, this element in turn decays into one with 94 protons and 94 electrons. This element, named plutonium, is radioactive but long-lived. The theoretical work of two British scientists suggested that plutonium would be an even better source of fission than U-235. In that case, costly separation of isotopes might not be necessary; the operation of a reactor fueled by U-235 might actually produce additional fuel from previously unusable but available U-238. No one was to guess that construction of larger plutonium breeder reactors would produce as many problems and delays as have since occurred.

11. Groves, *Now*, p. 180. Fletcher Knebel and Charles W. Bailey II, *No High Ground* (New York: Harper, 1960), pp. 65-66.

12. Hewlett and Anderson, *New World*, p. 285. The Military Policy Committee was formed in September 1942 as a sort of board of directors with army, navy, and civilian science representatives to whom Groves, a military man, would report as well as to the civilian S-1 committee. The Military Policy Committee was chaired by Dr. Vannevar Bush, with Dr. James Conant as his alternate.

13. Groves, *Now*, p. 179. Note of meeting of 8 May 1944, MMRD, H-B, 54. Gowing, *Britain and Atomic Energy*, p. 298.

14. Diplomatic history of negotiations prepared by Groves for Lovett, MMRD, H-B, 111.

15. Gowing, *Britain and Atomic Energy*, pp.182-85. Bothwell, *Eldorado*, pp. 110-11, 126. See also John Wheeler-Bennett, *Sir John Anderson, Viscount Waverley* (New York: St. Martin's Press, 1962), p. 289.

16. John Akers, as quoted in Bothwell, *Eldorado*, p. 140. See also Eggleston, *Canada's Nuclear Story*, pp. 81-82.

17. Gowing, *Britain and Atomic Energy*, p. 298.

18. Minutes of meeting of CPC, 17 Dec. 1943, MMRD, H-B, 10.

CHAPTER 2: THE CORNERSTONE: AGREEMENT WITH BELGIUM

1. Campbell to Bundy, 24 Feb. 1944, MMRD, H-B, 54. Groves, *Now*, p. 171. Much of this chapter originally appeared as "The Uranium Negotiations of 1944" in *Le Congo Belge durant la Seconde Guerre Mondiale* (Brussels: Académie royale des sciences d'outre-mer/Koninklijke Academie voor Overzeese Wetenschappen, 1983), pp. 253-83. The author is grateful to the Academy for its permission to reuse this material.

2. Anderson to Winant, 23 March 1944, MMRD, H-B, 54.

3. Anderson to Winant, 23 March 1944, ibid.

4. Camille Gutt, *La Belgique au Carrefour, 1940-1944* (Paris: Fayard, 1971), p. 166.

5. Note of conversation, 27 March 1944, MMRD, H-B, 54. The Americans thought 5,000 tons of ore could be produced fairly quickly were the Shinkolobwe mine reopened and electricity provided. The British reported that oxide had been selling at seven to fifteen shillings per pound in England before the war.

6. Note of meeting 27 March 1944, ibid.

7. Groves, *Now*, p. 175.

8. Note of meeting 8 May 1944, MMRD, H-B, 54.

9. Ibid., and memorandum regarding meeting of 8 May 1944.

10. Ibid. and report to chancellor by Barnes, Gorell Barnes, Betts and Traynor, 10 May 1944. Given the American presence at Shinkolobwe in 1943, the diplomats' desire to keep Congo authorities uninformed is amusing; Winant and Anderson of course did not know of the mission until the 8 May meeting.

11. Report to the chancellor by Barnes, Gorell Barnes, Betts and Traynor, 10 May, MMRD, H-B, 54.

12. Ibid.

13. Ibid.

14. Note of meeting of 11 May 1944, ibid.

15. Draft treaty (drawn by Sir Thomas Barnes), ibid., 35.

16. Betts to Stimson, 17 May 1944; ibid.; see also Anderson to Dill and Campbell, 17 May 1944, ibid., 17.

17. Draft treaty, ibid., 35. The significant "and strategic" words, which gave this clause broader scope, were inserted by Betts as an amendment.

18. Chronological history of negotiations, 25 Aug. 1944; Sengier to Hambro, 5 Sept. 1944, ibid., 35, 37.

19. Bundy memorandum regarding his July 1944 trip to U.K. describes the meetings of 13 and 16 July, ibid., 16.

20. Ibid.; also note of meeting 19 July 1944, ibid., 35.

21. Note of meeting 19 July 1944, ibid., 35.

22. Ibid.

23. Note of meeting 28 July 1944, ibid. The idea of offering neutralized rather than untreated ore had been reported at the 19 July meeting. The change of position may have been stimulated by comments from Groves.

24. Ibid.

25. Anderson to Campbell, 31 July 1944 (CANAM 99), ibid., 17.

26. Anderson to Campbell, 31 July 1944 (CANAM 100), ibid., 17.

27. Bundy to Peabody, 13 Aug. 1944, ibid., 35.

28. Draft agreement prepared by de Vleeschauwer, 14 Aug. 1944, ibid.

29. Gorell Barnes to Campbell, 17 Aug. 1944, ibid., 17.

30. Note of meeting 16 Aug. 1944, ibid., 35.

31. Ibid.

32. Ibid.

33. Ibid.

34. Ibid.; also Gorell Barnes to Campbell, 17 Aug. 1944, and chronological history of negotiations, 25 Aug. 1944, ibid., 17, 35.

35. Paul-Henri Spaak, *Combats inachevés*, 2 vols. (Paris: Fayard, 1969), 1:132.

36. Office of Strategic Services, Survey of the Belgian Congo, Sept. 1942, p. 35, U.S. Military History Institute Archives (Carlisle, PA), William Donovan papers, 19.

37. Stimson to Roosevelt, 25 Aug. 1944; Gorell Barnes to Campbell, 17 Aug. 1944 (CANAM 108, 109, 110), MMRD, H-B, 35, 17.

38. Bundy to Winant (?), 2 Sept. 1944, ibid.; see also Gorell Barnes to Campbell, 29 Aug. 1944, ibid.

39. Winant to Bundy, 6 Sept. 1944; Bundy to Winant, 4 Sept. 1944, ibid.

40. Winant to Bundy, 6 Sept. 1944, ibid.

41. The Memorandum of Agreement between the United States, the United Kingdom, and Belgium Regarding Control of Uranium is printed in *Foreign Relations of the United States* (Washington: U. S. Government Printing Office; hereafter cited as *FRUS*), *1944*, 2:1029-30. At the time of its signing, Article 10 required that it be "treated as a military secret in keeping with its purpose."

42. Ibid.

43. Winant to Bundy, 6 Sept. 1944; Betts to Bundy, 6 Sept. 1944; Bundy to Winant, 14 Sept. 1944, MMRD, H-B, 35.

44. Winant to Bundy, 6 Sept. 1944, ibid.

45. Gorell Barnes to Hambro, 19 Sept. 1944; Gorell Barnes to Campbell and Hambro, 19 Sept. 1944, ibid., 17.

46. Bundy to Stevenson, 18 Sept. 1944, marked "cancelled" in pencil, ibid., 5.

47. Sengier to Hambro, 5 Sept. 1944, ibid., 57.

48. Ibid.

49. P.L.G. to Groves, 8 Sept. 1944, ibid.

50. Memorandum of changes to draft, undated, ibid.

51. Contract between CDT and African Metals Corporation; see also Sengier to Hambro, 5 Sept. 1944, ibid.

52. Gorell Barnes to Dill and Campbell, 4 Oct. 1944; same to same, 11 Oct. 1944, ibid., 65. Also Groves to Patterson, 2 Dec. 1945, *FRUS, 1945,* 2:81. The covering letters from Spaak to Winant, 26 Sept. 1944, and from Winant to Spaak, 26 Sept. 1944, appear in *FRUS, 1944,* 2:1028, 1030. These are the same letters drafted by Gorell Barnes on 17 August but not used at that time.

53. United States Atomic Energy Commission, *In the Matter of J. Robert Oppenheimer* (Washington: U.S. Government Printing Office, 1954), p. 173.

54. Annette Baker Fox,*The Power of Small States: Diplomacy in World War II* (Chicago: Univ. of Chicago Press, 1959), p. 3.

55. See Jonathan E. Helmreich, *Belgium and Europe: A Study in Small Power Diplomacy* (The Hague: Mouton, 1976), pp. 405-11.

56. Kerr (British Ambassador to the Soviet Union) to Bevin, 3 Dec. 1945, *FRUS, 1945,* 2:82-84.

CHAPTER 3: EFFORTS AT PREEMPTION: BRAZIL, THE
NETHERLANDS, AND SWEDEN

1. Gowing, *Britain and Atomic Energy,* p. 180.

2. Anthony Cave Brown and Charles B. MacDonald, eds., *The Secret History of the Atomic Bomb* (New York: Dial Press/James Waid, 1977), pp. 191-93. Groves, *Now,* pp. 180-81. Bothwell, *Eldorado,* pp. 148-49.

3. Campbell to secretaries of CPC, 24 Aug. 1944, MMRD, H-B, 103.

4. Groves, *Now,* pp. 182-84. Brown and MacDonald, *Secret History,* p. 192. Gowing, *Britain and Atomic Energy,* pp. 303-4. Experiments in 1942 indicated that thorium, with 90 protons and 142 neutrons, when subjected to bombardment in a cyclotron would decay into a fissionable 233 isotope of uranium. The United States did not have a plant or technology for extracting U-233 from thorium efficiently, and so little attention was at first paid to obtaining monazite. Then in April 1944 further experimentation suggested that, once reaction had been initiated in a thorium pile with the aid of uranium, the reaction and the production of U-233 could be continued with the addition only of more thorium. Thus thorium, at that time considered ten times more plentiful than uranium, immediately became of strategic value. Hewlett and Anderson, *New World,* pp. 286-87.

5. A Survey of the World's Sources of Uranium and Thorium, Section I, enclosed in Groves to Stimson, 24 Nov. 1944, MMRD, H-B, 27.

6. Ibid.

7. Ibid.

8. Ibid.

9. Ruhoff to Groves, 27 Dec. 1943, MMRD, H-B, 68.

10. Merritt memorandum regarding visit to Eldorado mine, Northwest Territory, 22-25 March 1944, MMRD, Manhattan Engineering District Decimal Files 1942-48, 410.2, 68.

11. Groves to Stimson, 24 Nov. 1944, MMRD, H-B, 27.

12. Ibid.

13. Ibid.

14. Ibid.

15. Groves to Stimson, 9 Dec. 1944, MMRD, H-B, 5. Gowing, *Britain and Atomic Energy*, p. 317.

16. Groves to Bundy, 6 Feb. 1945, MMRD, H-B, 27.

17. Ibid.

18. Ibid.

19. Telephone interview with Joseph Volpe, 10 July 1984.

20. Sources and Supplies of Thorium, 9 May 1944, report enclosed in W. Gorell Barnes to Traynor, 17 May 1944, MMRD, Manhattan Engineering District Top Secret Files 1942-46, 16.

21. Ibid.

22. Groves, *Now*, p. 184. Groves memorandum, 23 Feb. 1945, *FRUS, 1945*, 2:5-7. There is an unaddressed working memorandum in MMRD, H-B, 34, dated 12 Feb. 1945, stating that certain unidentified individuals recognize the force of "your price point." Because of uncertainty as to the date of any purchase, "it is difficult to try to tie _____ down to anything more precise than broad understanding that if and when time comes he will deal reasonably and not seek to inflate price." A promise is made to get the message to someone who would do his best to cover the point in his discussion.

23. Groves memorandum, 23 Feb. 1945, *FRUS, 1945*, 2:5-7. War department report on Bouças, 19 Feb. 1945, MMRD, H-B, 34.

24. Groves memorandum, 23 Feb. 1945, *FRUS, 1945*, 2:5-7. War department report on Bouças, 19 Feb.1945, MMRD, H-B, 34.

25. Groves to Stimson, 18 June 1945, MMRD, H-B, 34. Telephone interview with John Lansdale, Jr., 30 June 1984.

26. McAshan and Lansdale to Groves, 10 July 1945, *FRUS, 1945*, 2:14-19.

27. Memorandum of Agreement between the United States of Brazil and the United States of America enclosed in Berle to Byrnes, 10 July 1949, *FRUS, 1945*, 2:19-23.

28. McAshan and Lansdale to Groves, 10 July 1945, *FRUS, 1945*, 2:14-19.

29. Memorandum of Agreement between the United States of Brazil and the United States of America enclosed in Berle to Byrnes, 10 July 1945, *FRUS, 1945*, 2:19-23.

30. McAshan to Clayton, 12 Sept. 1945, MMRD, H-B, 34. Telephone interview with John Lansdale, Jr., 30 June 1984.

31. McAshan and Lansdale to Groves, 10 July 1945, *FRUS, 1945*, 2:14-19.

32. Ibid. Also telephone interview with John Lansdale, Jr., 30 June 1984.

33. Minutes of meeting of CPC, 4 July 1945, MMRD, H-B, 37.

34. See Halifax to Acheson, 24 Sept. 1945, *FRUS, 1945*, 2:47-48.

35. Minutes of meeting of CPC, 4 July 1945, MMRD, H-B, 37.

36. Groves to Stimson, 7 June 1945, MMRD, H-B, 37. Bundy record of conversations with British representatives, entry for 19 June 1945, MMRD, H-B, 52.

37. Traynor memorandum, 3 Aug. 1945, *FRUS, 1945,* 2:25-36.

38. Ibid. Telephone interview with John Lansdale, Jr., 30 June 1984.

39. Groves to Patterson, 13 Oct. 1945, MMRD, H-B, 38.

40. These last points are inferred from references made in a British memorandum regarding negotiations with Sweden sent to Washington legation, 15 Aug. 1945, MMRD, H-B, 53.

41. Diplomatic History of Manhattan Project, p. 28, MMRD, H-B, 111.

42. Ibid., p. 29. Makins memorandum, 2 July 1945, MMRD, H-B, 37. Bundy record of conversations with British representatives, entry for 1 July 1945, MMRD, H-B, 52. Diplomatic History of Manhattan Project, note by Groves p. 29, MMRD, H-B, 111.

43. Makins memorandum, 2 July 1945, MMRD, H-B, 37.

44. Ibid.

45. Minutes of meeting of CPC, 4 July 1945, MMRD, H-B, 37.

46. Diplomatic History of Manhattan Project, note by Groves p. 29, MMRD, H-B, 111.

47. Traynor memorandum, 3 Aug. 1945, *FRUS, 1945,* 2:25-36.

48. Heads for Proposed Tripartite Agreement, 18 July 1945, MMRD, H-B, 157.

49. Ibid.

50. Ibid.

51. Telephone interview with John Lansdale, Jr., 30 June 1984.

52. Johnson memorandum, 10 Aug. 1945, *FRUS, 1945,* 2:37-40.

53. British memorandum regarding negotiations with Sweden sent to Washington legation, 15 Aug. 1945, MMRD, H-B, 53.

54. Draft agreement presented to Swedish government, MMRD, H-B, 53.

55. Ibid.

56. British memorandum regarding negotiations with Sweden sent to Washington legation, 15 Aug. 1945, MMRD, H-B, 53.

57. Ibid.

58. Ibid. Diplomatic History of Manhattan Project, p. 30, MMRD, H-B, 111.

59. As quoted in Johnson memorandum, 10 Aug. 1945, *FRUS, 1945,* 2:37-40.

60. Ibid.

61. Diplomatic History of Manhattan Project, p. 29, MMRD, H-B, 111. Groves to Patterson, 13 Oct. 1945, MMRD, H-B, 38.

62. Vance to Groves, 25 Sept. 1945, *FRUS, 1945,* 2:50-53.

63. Ibid.

64. Telephone interviews with John Lansdale, Jr., 30 June 1984, and Joseph Volpe, 10 July 1984.

65. Undén to Johnson, 11 Sept. 1945, enclosed in Johnson to Groves, 22 Sept. 1945, *FRUS, 1945,* 2:45-47.

66. Ibid.

67. Johnson to Groves, 22 Sept.1945, *FRUS, 1945*, 2:45-46.

68. Minutes of meeting of CPC, 13 Oct. 1945, MMRD, H-B, 38.

69. Gowing, *Britain and Atomic Energy*, pp. 313-14. Minutes of meeting of CPC, 4 Dec. 1945, *FRUS, 1945*, 2:86-89.

70. Groves to Patterson, 3 Dec.1945, *FRUS, 1945*, 2:84-85.

71. Gowing, *Britain and Atomic Energy*, p. 314. Groves, *Now*, pp. 182-83.

72. U.S. Atomic Energy Commission, *Oppenheimer*, p. 176. Nikolai Grishin, "The Saxony Uranium Mining Operation ('Vismut')," in *Soviet Economic Policy in Postwar Germany: A Collection of Papers by Former Soviet Officials*, ed. by Robert Slusser (New York: Research Program on the U.S.S.R., 1953), pp. 127-53.

CHAPTER 4: PRICE, POLITICS, AND PRIDE: FURTHER
NEGOTIATIONS WITH BELGIUM

1. Memorandum of Agreement between the United States, the United Kingdom and Belgium Regarding Control of Uranium, *FRUS, 1944*, 2:1029-30.

2. Minutes of meeting of CPC, 8 March 1945, *FRUS, 1945*, 2:7-11.

3. Groves to Patterson, 2 Dec. 1945, *FRUS, 1945*, 2:81-82.

4. Ibid., and minutes of meeting of CPC, 4 Dec. 1945, *FRUS, 1945*, 2:86-89.

5. Kirk to Byrnes, 15 May 1946, U. S. National Archives, Department of State Records (hereafter cited as Nat. Arch., State), 855.64615-1546.

6. Robert Murphy, *Diplomat*, p. 329.

7. Acheson to Kirk, 11 Oct. 1946, Nat. Arch., State, 855.6359/10-1146.

8. Kirk to Byrnes, 8 Oct. 1946, ibid., 855.6359/10-846.

9. Kirk to Acheson, 15 Oct. 1946, ibid., 855.646/10-1546.

10. Kirk to Acheson, 30 Oct. 1946, and same to same, 1 Nov. 1946, ibid., 855.646/10-3046, 855.6359/11-146.

11. Kirk to Acheson, 10 May 1947, ibid., 855A.6359/5-1247.

12. Achilles to Acheson, 16 June 1947; same to same, 24 June 1947, ibid., 855A.6359/6-1647, 855A.6359/6-2447.

13. Achilles to Acheson, 24 June 1947, ibid., 855.646/6-2447.

14. Ibid.

15. Minutes of meeting of CPC, 4 Dec. 1945, *FRUS, 1945*, 2:86-89.

16. Patterson to Byrnes, 22 Dec. 1945, Nat. Arch., State, 855.646/12-2245.

17. Patterson to Byrnes, 5 Jan. 1946, and same to same, 25 Jan. 1946, ibid., 855.646/1-546, 855.646/1-2546.

18. Kirk to Byrnes, 2 April 1945, ibid., 855.646/4-346.

19. Kirk to Byrnes, 8 April 1946, ibid., 855.646/4-846.

20. Kirk to Byrnes, 9 April 1946, ibid., 855.646/4-946.

21. Kirk to Byrnes, 18 April 1946, *FRUS, 1946*, 1:1233-34.

22. Ibid.

23. Groves memorandum, 1 April 1946, U. S. National Archives, Modern

Military Records Division, Manhattan Engineering District Top Secret Files, 20 Tab. V.

24. Groves to Acheson, 29 April 1946, *FRUS, 1946,* 1:1240-41.

25. Groves to Acheson, 29 April 1946, *FRUS, 1946,* 1:1237-38. The use of the word "political" is of interest, for Americans and Belgians alike had been careful in 1944 not to so describe the Tripartite Agreement in order to avoid constitutional requirements for setting it before their legislative bodies.

26. Ibid.

27. Acheson to Kirk, 2 May 1946, *FRUS, 1946,* 1:1243-44. Kirk to Byrnes, 5 May 1946, Nat. Arch., State, 855.646/5-546.

28. Kirk to Byrnes, 15 May 1946, Nat. Arch. State, 855.646/5-1546.

29. Kirk to Byrnes, 18 May 1946, 1:1247-48.

30. Ibid. and editor's footnote; Kirk to Byrnes, 24 July 1946; Kirk to Byrnes, 9 Aug. 1946; Acheson to Kirk, 28 Aug. 1946, Nat. Arch., State, 855.646/7-2446, 855.646/8-946, 855.646/8-946.

31. Kirk to Byrnes, 10 Jan. 1947, *FRUS, 1947,* 1:783-84.

32. Kirk to Acheson, 17 Jan. 1947, Nat. Arch., State, 855.6359/1-1747.

33. Marshall to Kirk, 10 Feb. 1947, *FRUS, 1947,* 1:793-94.

34. Kirk to Byrnes, 10 Jan. 1947, *FRUS, 1947,* 1:783-84.

35. Kirk to Acheson, 5 Feb. 1947, verbatim text (translation) of Spaak memorandum of 4 Feb. 1947, *FRUS, 1947,* 1:792-93.

36. Marshall to Kirk, 10 Feb. 1947, *FRUS, 1947,* 1:793-94.

37. Memorandum by Gullion of conversation with Donald P. Maclean, 18 Feb. 1947; Marshall to Kirk, 10 Feb. 1947, *FRUS, 1947,* 1:793-94 and editor's note including extract from British message of 31 Jan.

38. Gullion to Acheson, 26 March 1947, *FRUS, 1947,* 1:802-4.

39. Ibid.

40. Ibid.

41. Kirk to Marshall, 13 March 1947, Nat. Arch., State, 855.646/3-1347.

42. Kirk to Acheson, 10 Jan. 1947, ibid., 855.646/1-1047.

43. Kirk to Acheson, 25 Feb. 1947, *FRUS, 1947,* 1:796-97.

44. Achilles to Acheson, 18 March 1947, Nat. Arch., State, 855.6359/3-1847. This last was a change from the position taken by Spaak in mid-January. Kirk to Acheson, 17 Jan. 1947, ibid., 855.6359/1-1747.

45. Kirk to Acheson, 8 Nov. 1946, ibid., 855.646/11-846.

46. Gullion to Acheson, 26 March 1947, *FRUS, 1947,* 1:802-4.

47. Ibid.

48. Kirk to Acheson, 15 April 1947, Nat. Arch., State, 855.646/4-1547.

49. Kirk to Acheson, 14 May 1947, verbatim text translation, *FRUS, 1947,* 1:812.

50. Marshall (marked drafted and sent by Acheson) to Kirk, 14 May 1947, *FRUS, 1947,* 1:812-15.

51. Ibid.

52. Marshall to Kirk, 6 May 1947, Nat. Arch., State, 855.6359/1-1747.

53. Achilles to Marshall, 19 May 1947, *FRUS, 1947,* 1:815-17.

54. Achilles to Acheson, 27 May 1947, Nat. Arch., State, 855A.6359/5-2747. He urged creating a Belgian sense of dependence on U.S. research.

55. Achilles to Marshall, 19 May 1947, *FRUS, 1947*, 1:815-17.

56. Memorandum by Gullion of conversation and meeting between Atomic Energy Commission and Kirk, 4 June 1947, *FRUS, 1947*, 1:818-22.

57. Acheson to Kirk, 27 June 1947, *FRUS, 1947*, 1:822-24.

58. Kirk to Marshall, 3 July 1947, *FRUS, 1947*, 1:824.

59. Kirk to Marshall, 4 July 1947, *FRUS, 1947*, 1:825.

60. Acheson to Kirk, 27 June 1947, *FRUS, 1947*, 1:822-24.

61. Marshall to Clayton, 24 July 1947, *FRUS, 1947*, 1:828-29.

62. Kirk to Lovett, 2 Sept. 1947, *FRUS, 1947*, 1:835-37.

63. Kirk to Acheson, 2 July 1947; Kirk to Acheson, 24 June 1947, Nat. Arch., State, 855A.6359/7-247, 855A.6359/6-2447. 1:828-29.

64. Kennan to Lovett, 24 Oct. 1947, and editor's footnote including abstracts from AEC letter of 1 Oct. 1947, *FRUS, 1947*, 1:842-43.

65. Kirk to Lovett, 2 Sept. 1947, *FRUS, 1947*, 1:835-37.

66. Ibid.

67. Minutes of meeting of secretaries of state, war, and navy, 11 Sept. 1947, *FRUS, 1947*, 1:838-40.

68. Forrestal memorandum of conversation, 16 Nov. 1947, *FRUS, 1947*, 1:864-66.

69. Minutes of meeting, 26 Nov.1947, *FRUS, 1947*, 1:870-79.

70. Paraphrase of British foreign office telegram, 25 Sept. 1947, *FRUS, 1947*, 1:840.

71. Achilles (now stationed in Washington with the state department's Division of Western European Affairs) memorandum of conversation, 30 Oct. 1947, *FRUS, 1947*, 1:841-42.

72. Hugh Millard (new chargé in Brussels) to Marshall, 24 Nov. 1947, Nat. Arch., State, 855A.6359/11-2447; Gullion to embassy in Belgium, 1 Nov. 1947, *FRUS, 1947*, 1:851-52.

73. Gullion to embassy,1 Nov.1947, *FRUS, 1947*, 1:851-52.

74. AEC letter of 1 Oct. 1947 quoted in footnote to Kennan memorandum to Lovett, 24 Oct. 1947, *FRUS, 1947*, 1:843.

75. Millard to Lovett, 24 Dec. 1947, Nat. Arch., State, 855.6359/12-2447; Silvercruys to Marshall, 19 Jan. 1948, *FRUS, 1948*, 1:pt. 2, 687-88.

76. Millard to Marshall, 7 Feb. 1948, and editor's note, *FRUS, 1948*, 1:pt. 2, 691-92.

77. Marshall to embassy, 9 March 1948, *FRUS, 1948*, 1:pt. 2, 693-94.

78. Note by American members of CPC, 6 July 1948, *FRUS, 1948*, 1:pt. 2, 726-27.

79. Kirk to Marshall, 11 July 1948, Nat. Arch., State, 855.646/7-1148.

80. Ibid. Also Marshall to embassy, 16 July 1948; Kirk to Marshall, 2 Aug. 1948, Nat. Arch., State, 855A.6359/7-948, 855.646/8-248.

81. Lovett to Kirk, 17 Sept. 1948, Nat. Arch., State, 855A.6359/9-1748; memorandum of conversation, 25 Aug. 1948, *FRUS, 1948*, 1:pt. 2, 747-48.

82. Hewlett and Anderson, *Atomic Shield*, p. 174. Murphy to Acheson, 10 Jan. 1950, *FRUS, 1950*, 1:493-99.

CHAPTER 5: RELUCTANT ANGLO-AMERICAN COLLABORATION

1. Gowing, *Britain and Atomic Energy*, pp. 123-26. Hewlett and Anderson, *New World*, p. 259. David E. Lilienthal, *The Journals of David E. Lilienthal*, vol. 2: *The Atomic Energy Years* (New York: Harper and Row, 1964), entry for 9 Feb. 1949, p. 465.

2. Gowing, *Britain and Atomic Energy*, pp. 155, 160. Hewlett and Anderson, *New World*, p. 268.

3. Gowing, *Britain and Atomic Energy*, pp. 162-64.

4. Ibid., p. 168. The Articles of Agreement governing collaboration between the authorities of the U.S.A. and the U.K. in the matter of Tube Alloys may be found in U.S. Department of State, *Collaboration in Atomic Energy Research and Development*, Publication 5561 (Washington: U.S. Government Printing Office, 1954).

5. Hewlett and Anderson, *New World*, p. 282.

6. As quoted in Eggleston, *Canada's Nuclear Story*, p.93.

7. Hewlett and Anderson, *New World*, p. 228.

8. See MMRD, H-B, 36; also Martin J. Sherwin, *A World Destroyed: The Atomic Bomb and the Grand Alliance* (New York: Knopf, 1975), pp. 135-36.

9. Hewlett and Anderson, *New World*, p. 336.

10. Gowing, *Britain and Atomic Energy*, pp. 196-99, 295.

11. Minutes of meeting of CPC, 4 July 1945, *FRUS, 1945*, 2:12-14.

12. McMahon Bill, Hewlett and Anderson, *New World*, pp. 714-22.

13. U.S. Department of State, *Collaboration*; also, *FRUS, 1945*, 2:64fn.

14. Bush to Byrnes, 15 Nov. 1945, *FRUS, 1945*, 2:69-73.

15. Gowing, *Independence* 1:57. Wheeler-Bennett, *Sir John Anderson*, pp. 333-36 and *passim*.

16. Arneson to Patterson, 17 April 1946, *FRUS, 1945*, 2:63-69.

17. Interim Committee log, entry for 15 Nov.1945, MMRD, H-B, 98.

18. Hewlett and Anderson, *New World*, p. 468. Arneson to Patterson, 17 April, 1946; memorandum by Truman, Attlee, King, 16 Nov. 1945; Groves and Anderson to Patterson, 16 Nov. 1945, *FRUS, 1945*, 2:63-69, 75-76.

19. Groves and Arneson to Patterson, 16 Nov. 1945, *FRUS, 1945*, 2:75-76.

20. Minutes of meeting of CPC, 4 Dec. 1945, *FRUS, 1945*, 286-89.

21. Memorandum by Groves, 2 Jan. 1946, *FRUS, 1946*, 1:1197-1203.

22. Telephone interview with Joseph Volpe, 10 July 1984.

23. Lilienthal, *Journals*, entry for 16 Jan. 1946, 2:10-11.

24. Draft report, 15 Feb. 1946, *FRUS, 1946*, 1:1207-13. Hewlett and Anderson, *New World*, p. 478. Gowing, *Independence*, 1:96.

25. Groves to Byrnes, 13 Feb. 1946, *FRUS, 1946*, 1:1204-07.

26. Lilienthal, *Journals*, entry for 6 March 1946, 2:25-26.

27. U.S. Atomic Energy Commission, *Oppenheimer*, p. 175.

28. Lilienthal, *Journals*, entry for 6 March 1946, 2:25-26.

29. Gowing, *Independence*, 1:100. Memorandum by British members of CPC, undated, *FRUS, 1946*, 1:1225-27.

30. Suggested draft by U. S. members of CPC, 9 April 1946; Groves to Acheson and Bush, 29 April 1946, *FRUS, 1946*, 1:1226-27, 1238-40.

31. Minutes of meeting of CPC, 15 April 1946, *FRUS, 1946,* 1:1227-31.

32. Attlee to Truman, 16 April 1946, *FRUS, 1946,* 1:1231-32.

33. Truman to Attlee, 20 April, 1946; Acheson to Truman, 6 May 1946, *FRUS, 1946,* 1235-37, 1244-45. Bothwell, *Eldorado,* p. 179.

34. Groves to Acheson, 29 April 1946, *FRUS, 1946,* 1:1240-41.

35. Ibid.

36. Ibid.

37. Gowing, *Independence,* 1:103.

38. Memorandum by Hancock, 1 May 1946, *FRUS, 1946,* 1:1242-43.

39. Memorandum by Acheson, Bush, Groves to U.S. members of CPC, 7 May 1946; memorandum of subcommittee to CPC, 13 May 1946, *FRUS, 1946,* 1:1245-57.

40. Memorandum of Groves, Makins, Bateman, 26 July 1946; minutes of meeting of CPC, 31 July 1946, *FRUS, 1946,* 1:1254-57.

41. Gowing, *Independence,* 1:356.

42. Groves to Patterson, 3 Dec. 1945, MMRD, H-B, 40.

43. Dean Acheson, *Present at the Creation: My Years at the State Department* (New York: Norton, 1969), p. 167. For the view of many scientists on the legislation, see Alice Kimball Smith, *A Peril and a Hope; the Scientists' Movement in America: 1945-47* (Chicago: Univ. of Chicago Press, 1965). The legislation, Public Law 585, may be found in *United States Statutes at Large,* 79th cong., 2nd session, 1946 (Washington: U.S. Government Printing Office, 1947), 60:pt. 2, 755-75.

44. Gowing, *Independence,* 1:109.

45. Attlee to Truman, 7 June 1946, *FRUS, 1946,* 1:1249-53.

46. Gowing, *Independence,* 1:110.

47. Kenneth Harris, *Attlee* (New York: Norton, 1982), p. 286. Gowing, *Independence,* 1:119.

48. Acheson memorandum of conversation, 1 Feb. 1947, *FRUS, 1947,* 1:785-89.

49. Lilienthal, *Journals,* entry for 5 July 1947, 2:219.

50. Ibid., entry for 21 July 1947, 2:228.

51. Ibid. and entry for 11 March 1948, 2:303. Also Hewlett and Duncan, *Atomic Shield,* p. 152.

52. Acheson statement to Joint Congressional Committee, 12 May 1947, *FRUS, 1947,* 1:806-11.

53. Hickenlooper to Marshall, 29 Aug. 1947, *FRUS, 1947,* 1:833-34.

54. Minutes of meeting of secretaries of state, war, and navy, 11 Sept. 1947, *FRUS, 1947,* 1:838-40.

55. Patterson and Forrestal to Marshall, undated, *FRUS, 1947,* 1:798-99.

56. Memorandum by Kennan and Gullion, 24 Oct. 1947, *FRUS, 1947,* 1:844-47.

57. Minutes of meeting of American members of CPC, 5 Nov. 1947, *FRUS, 1947,* 1:852-60.

58. Lilienthal, *Journals,* entry for 22 Nov. 1947, 2:259.

59. Hewlett and Duncan, *Atomic Shield,* p. 277. Minutes of meeting of American members of CPC, 24 Nov. 1947, *FRUS, 1947,* 1:866-70.

60. Minutes of meeting of American members of CPC with chairmen of Joint Congressional Committee on Atomic Energy and Senate Foreign Relations Committee, 26 Nov. 1947, *FRUS, 1947,* 1:870-79. Also, Foster memorandum 26 Feb. 1947, ibid., 1:797-98.

61. Minutes of meeting of American members of CPC with chairmen of Joint Committee and Senate Foreign Relations Committe, 26 Nov. 1947, *FRUS, 1947,* 1:870-79.

62. Lovett to Douglas, 4 Dec. 1947, *FRUS, 1947,* 1:879-81.

63. Minutes of meeting of CPC, 10 Dec. 1947; memorandum of subgroup on technical cooperation, 12 Dec. 1947, *FRUS, 1947,* 1:889-96.

64. Minutes of meeting of CPC, 15 Dec. 1947, *FRUS, 1947,* 1:897-903.

65. Hewlett and Duncan, *Atomic Shield,* p. 282.

66. Record of Kennan-Bohlen teletype conference, 17 Dec. 1947, *FRUS, 1947,* 1:905-6.

67. Ibid.

68. Gowing, *Independence,* 1:248-51.

69. As quoted ibid., 1:252.

70. Gullion to Lovett, 20 Dec. 1947, *FRUS, 1947,* 1:907-8.

71. Gowing, *Independence,* 1:107.

72. Ibid., 1:254.

73. Ibid., 1:363.

CHAPTER 6: THE DIFFICULTIES OF SHARING

1. Margaret Gowing, *Independence,* 1:255-56.

2. Hewlett and Duncan, *Atomic Shield,* p. 152. Gullion to Lovett, 19 March 1948, *FRUS, 1948,* 1:pt. 2, 700.

3. Carpenter to Forrestal, 12 Aug. 1948; Bush to Forrestal, 12 Aug. 1948, *FRUS, 1948,* 1:pt. 2, 734-39.

4. Gowing, *Independence,* 1:258-60.

5. Moore to Forrestal, 1 Sept. 1948, *FRUS, 1948,* 1:pt. 2, 750-52.

6. Carpenter to Forrestal, 16 Sept. 1948; Arneson memorandum, 27 Sept. 1948; Wendel to Lovett, 30 Sept. 1948, *FRUS, 1948,* 1:pt. 2, 755-58, 767-72. Gowing, *Independence,* 1:260.

7. Arneson to Lovett, 2 Nov. 1948, *FRUS, 1948,* 1:pt. 2, 781-84.

8. Ibid.

9. Arneson to Acheson and Webb, 3 Feb. 1949, *FRUS, 1949,* 1:419-28. Also Hewlett and Duncan, *Atomic Shield,* p. 276.

10. Souers to Truman with annexed report, 2 March 1949, *FRUS, 1949,* 1:443-61.

11. Paper prepared by Working Group of Special Committee, 1 March 1949 annexed to Arneson to Acheson, 1 March 1949, *FRUS, 1949,* 1:441-43.

12. Minutes of Special Committee, 2 March 1949, *FRUS, 1949,* 1:441-43.

13. Souers to Truman with annexed report, 2 March 1949, *FRUS, 1949,* 1:443-61.

14. Hewlett and Duncan, *Atomic Shield,* p. 304. Arneson memorandum, 6

July 1949; record of meeting at Blair House, 14 July 1949; record of meeting of Joint Committee, 20 July 1949; record of meeting of Joint Committee, 27 July 1949, *FRUS, 1949,* 1:471-73, 476-81, 490-98, 503-06.

15. Arneson to Webb, undated; same to same, 25 July 1949, *FRUS, 1949,* 1:513-14, 499-500.

16. Minutes of meeting of American members of the CPC, 13 Sept. 1949, *FRUS, 1949,* 1:520-26.

17. Lilienthal, *Journals,* entry for 19 July 1949, 2:548.

18. A foreign office member, as quoted by Gowing in *Independence,* 1:281.

19. Minutes of meeting of CPC, 20 Sept. 1949, *FRUS, 1949,* 1:529-35. Gowing, *Independence,* 1:285.

20. Webb memorandum of conversation with Truman, 1 Oct. 1949, *FRUS, 1949,* 1:552-53.

21. Memorandum for American members of CPC, 18 April 1950, annexed to minutes of meeting of American members of CPC, 25 April 1950, *FRUS, 1950,* 1:547-58.

22. Franks to Acheson, 29 Dec. 1949, *FRUS, 1949,* 1:620-22. Gowing, *Independence,* 1:297.

23. Fisher and Arneson to Acheson, 18 Jan. 1950, *FRUS, 1950,* 1:499-502. Lilienthal, *Journals,* entry for 30 Dec. 1949, 2:615.

24. Johnson to Acheson, 13 March 1950, Acheson to Johnson, 3 April 1950; Johnson to Truman, 24 Feb. 1950; report by Special Committee to Truman, 9 March 1950, *FRUS, 1950,* 1:541-42, 546-47, 538-39, 541-42. Hewlett and Duncan, *Atomic Shield,* p. 546.

25. Acheson memorandum, 16 May 1950, *FRUS, 1950,* 1:559-62.

26. Minutes of meeting of American members of CPC, 7 Sept. 1950, *FRUS, 1950,* 1:572-75.

27. Arneson to Acheson, 14 Dec. 1950, *FRUS, 1950,* 1:593-96.

28. Minutes of meeting of American members of CPC, 7 Sept. 1950, *FRUS, 1950,* 1:572-75.

29. Arneson to LeBaron, 9 Feb. 1951; Churchill to Truman, 12 Feb. 1951; Truman to Churchill, 24 March 1951, *FRUS, 1951,* 1:691-94, 703-4. The labor government was amenable to this quieting of what was viewed as a Churchillian publicity campaign.

30. As quoted in an AEC statement of 18 May 1959, annexed to Lay memorandum, 21 May 1951, *FRUS, 1951,* 1:721-30.

31. Dean to Foster, 27 Nov. 1951, *FRUS, 1951,* 1:785-88.

32. Statement of AEC, 18 May 1951, annexed to Lay memorandum, 21 May 1951, *FRUS, 1951,* 1:721-29.

33. Ibid.

34. Dean to Lay, 19 July 1951, *FRUS, 1951,* 1:747-48.

35. Minutes of meeting of American members of CPC, 24 Aug. 1951, *FRUS, 1951,* 1:755-63. Hewlett and Duncan, *Atomic Shield,* p. 481.

36. Minutes of meeting of American members of CPC, 24 Aug. 1951, *FRUS, 1951,* 1:755-63.

37. Ibid.

38. Dean to Foster, 27 Nov. 1951, *FRUS, 1951,* 1:785-88.

39. Minutes of meeting of CPC, 27 Aug. 1951, *FRUS, 1951,* 1:763-67.

40. Franks to Acheson, 26 Dec. 1951, *FRUS, 1951,* 1:798-99.

41. Minutes of meeting of American members of the CPC, 24 Aug. 1951, *FRUS, 1951,* 1:755-63.

42. Dean to Foster, 27 Nov. 1951, *FRUS, 1951,* 1:785-88.

43. Informal statement by department of defense, undated, *FRUS, 1951,* 1:769-72.

44. Ibid.

45. Lovett to Acheson, 12 Oct. 1951; Dean to Foster, 27 Nov. 1951, *FRUS, 1951,* 1:776-77, 785-88. Hewlett and Duncan, *Atomic Shield*, p. 483.

46. Foster to Dean, 2 Nov. 1951; Dean to Foster, 27 Nov. 1951, *FRUS, 1951,* 1:784-85, 785-88.

47. Lay report, 21 Dec. 1951, *FRUS, 1951,* 1:794-98.

48. Hewlett and Duncan, *Atomic Shield*, p. 575.

49. Arneson to Acheson, 16 Jan. 1952, *FRUS, 1952-54,* 2:pt. 2, 847.

50. Gowing, *Independence*, 1:415.

51. Brown and MacDonald, ed., *Secret History*, p. xvii. Memorandum of 16 Jan. 1952 meeting of special committee of NSC, 17 Jan. 1952, *FRUS, 1952-54,* 2:pt. 2, 851-58.

52. As quoted by Gowing, *Independence*, 1:265.

53. Acheson, *Present at the Creation*, p. 484. Gowing, *Independence*, 1:313.

54. Gowing, *Independence*, 1:413.

CHAPTER 7: MUCH EFFORT, LIMITED GAIN: CONTINUED
GLOBAL NEGOTIATIONS

1. Minutes of meeting of CPC, 4 Dec. 1945, MMRD, H-B, 40. Lovett to Marshall, 11 Aug. 1947, *FRUS, 1947,* 1:831-32.

2. Gullion to Acheson and annexed memorandum of Gullion conversation, 3 March 1947, *FRUS, 1947,* 1:799-800.

3. Ibid. Gullion to Lovett, 22 July 1947, *FRUS, 1947,* 1:827-28. Memorandum of Agreement between the United States of Brazil and the United States of America, enclosed in Berle to Byrnes, 10 July 1945, *FRUS, 1945,* 2:19-23.

4. Gullion to Lovett, 19 March 1948; Lovett to Marshall, 11 Aug. 1947; Lovett to embassy in Brazil, 29 Oct. 1947, *FRUS, 1947,* 1:827-28, 831-32, 847-48.

5. Gullion to Lovett, 19 March 1948; Key to Marshall, 13 April 1948, *FRUS, 1948,* 1:pt. 2, 700-701, 706-7.

6. Key to Marshall, 26 Jan. 1948; editorial note; Marshall to embassy in Brazil, 6 Feb. 1948; Key to Marshall, 30 March 1948; Harmon to Eakens, 19 Aug. 1948, *FRUS, 1948,* 9:352-54, 354-55, 359-60, 362-64.

7. Daniels to Thorp, 8 March 1948 and footnote; Clark memorandum of conversation, 24 March 1948; Marshall to embassy in Brazil, 3 Feb. 1948, *FRUS, 1948,* 9:377-78, 378-80, 398-99.

8. Marshall to embassy in Brazil, 9 July 1948; Johnson to Marshall, 15 July 1948; same to same 15 Oct. 1948; *FRUS, 1948,* 1:pt. 2, 727-28, 732-33, 775-76.

9. Johnson to Acheson, 14 April 1949; summary log of atomic energy work in the office of the under secretary of state, 1 Feb. 1949-31 Jan. 1950, *FRUS, 1949*, 1:463-64, 622-30.

10. U.S. embassy memorandum to Brazilian foreign office, 15 Dec. 1949, *FRUS, 1949*, 1:603. Johnson to Acheson, 24 March 1950, 1:547-58.

11. Memorandum by British members of CPC annexed to minutes of meeting of American members of CPC, 25 April 1950, *FRUS, 1950*, 1:547-58.

12. Arneson to Marten, 9 March 1951, *FRUS, 1951*, 1:696-99. See also summary log of atomic energy work in the office of the under secretary of state, May-Sept. 1950, *FRUS, 1950*, 1:580-88.

13. McFall to Javits, 4 April 1951, *FRUS, 1951*, 1:710-12.

14. Webb to embassy in Brazil, 6 Dec. 1951, *FRUS, 1951*, 1:790.

15. Johnson to Fontoura, 25 Oct. 1951; Johnson to Arneson, 11 Dec. 1951; Johnson to Acheson, 11 Dec. 1951, *FRUS, 1951*, 1:783-84, 790-92, 793.

16. Memorandum prepared in office of Under Secretary of State Lovett, 17 Sept. 1948, *FRUS, 1948*, 1:pt. 2, 758-65.

17. Ibid.

18. Ibid.

19. British embassy to department of state, 20 Oct. 1948, *FRUS, 1948*, 1:pt. 2, 776-79.

20. Memorandum of conversation, 21 Oct. 1948, *FRUS, 1948*, 1:pt. 2, 776-79.

21. Summary log of atomic energy work in the office of the under secretary of state, 1 Feb. 1949-31 Jan. 1950, *FRUS, 1950*, 1:622-30.

22. Henderson to Acheson, 29 July 1950, *FRUS, 1950*, 1:567-70.

23. Dean to Acheson, 19 Feb. 1951, *FRUS, 1951*, 1:694-95.

24. McFall to Javits, 4 April 1951, *FRUS, 1951*, 1:710-12.

25. Acting secretary of state to van Kleffens, 13 April 1948; Arneson memorandum of conversation, 31 Dec. 1948 and annexed memorandum of van Kleffens to acting secretary of state, 31 Dec. 1948, *FRUS, 1948*, 1:pt. 2, 704-6, 796-98.

26. Summary log of atomic energy work in the office of the under secretary of state, 1 Feb. 1949-31 Jan. 1950, *FRUS, 1949*, 1:622-30. Summary log of atomic energy work in the office of the under secretary of state, May-Sept. 1950, *FRUS, 1950*, 1:580-88.

27. Osborn to Rusk, 29 Oct. 1947, *FRUS, 1947*, 1:848-50.

28. Interdepartmental Intelligence Study, annexed to Hillenkoetter memorandum, 15 Dec. 1947, *FRUS, 1947*, 1:903-6. Uranium deposits in Saxony were not mentioned; Soviet exploitation of these, however, did not accelerate until 1948.

29. Gullion to Marshall, 9 March 1948, *FRUS, 1948*, 1:pt. 2, 699-700.

30. Report by National Security Council 3 Sept. 1948, *FRUS, 1948*, 3:232-34.

31. Hickerson to Lovett, 18 June 1948, *FRUS, 1948*, 1:pt. 2, 712-14.

32. Lovett to Matthews, 2 July 1948, *FRUS, 1948*, 1:pt. 2, 716-19.

33. Matthews to Lovett, 15 July 1948, *FRUS, 1948*, 1:pt. 2, 728-33.

34. Ibid.

35. Marshall to Matthews, 30 Aug. 1948, *FRUS, 1948*, 1:pt.2, 748-49.

36. British embassy to department of state, 16 Nov. 1948, *FRUS, 1948*, 1:pt. 2, 786-87. See also in the same volume, p. 746, the memorandum sent by the British embassy to the department of state, 20 Aug. 1948.

37. British embassy to department of state, 10 Nov. 1948; Lovett to Franks, 22 Nov. 1948, *FRUS, 1948*, 1:pt. 2, 786-87, 787.

38. Franks to Lovett, 21 Dec. 1948 and footnote quoting Lovett to Franks, 23 Dec. 1948, *FRUS, 1948*, 1:pt. 2, 792-93. Minutes of meeting of American members of CPC and annexes, 25 April 1950, *FRUS, 1950*, 1:547-58.

39. McKillop memorandum of conversation, 3 May 1949; summary log of atomic energy work in the office of the under secretary of state, 1 Feb. 1949-31 Jan. 1950, *FRUS, 1949*, 1:464-69, 622-30. Minutes of meeting of American members of CPC and annexes, 25 April 1950, *FRUS, 1950*, 1:547-58.

40. Perrin report of interview with Joliot, 5, 7 Sept. 1944, MMRD, H-B, 26.

41. Minutes of meeting of CPC, 22 Jan. 1945, *FRUS, 1945*, 1:2-5.

42. Gullion to Acheson, 7 March 1947, *FRUS, 1947*, 1:801-802.

43. Report to Truman by Special Committee of the National Security Council, 2 March 1949, *FRUS, 1949*, 1:443-61.

44. McKillop memorandum of conversation, 3 May 1949, *FRUS, 1949*, 1:464-69. Holmes to Acheson, 8 March 1950, *FRUS, 1950*, 1:541.

45. Chase memorandum of conversation, 21 Dec. 1949, *FRUS*, 1949, 1:617-20.

46. Summary log of atomic energy work in the office of the under secretary of state, 1 Feb. 1949-31 Jan. 1950, *FRUS, 1949*, 1:622-30.

47. Summary log of atomic energy work in the office of the under secretary of state, May-Sept. 1950; Murphy to Acheson, 10 Jan. 1950, *FRUS, 1950*, 1:580-88, 493-98.

48. Kirk to Acheson, 5 May 1950, *FRUS, 1950*, 1:559.

49. Bruce to Webb, 22 June 1950, *FRUS, 1950*, 1:564-65.

50. Terrill to department, 28 March 1951, *FRUS, 1951*, 1:704-9. See also Terrill to Arneson, 28 Dec. 1951, *FRUS, 1951*, 1:799-801.

51. Ibid.

52. Ibid.

53. Ibid.

54. Terrill to Arneson, 28 Dec. 1951, *FRUS, 1951*, 1:799-801.

55. Summary log of atomic energy work in the office of the under secretary of state, 1 Feb. 1949-31 Jan. 1950, *FRUS, 1949*, 1:622-30. See also Wilson to Marshall, 9 July 1947, *FRUS, 1947*, 1:825-27.

56. Summary log of atomic energy work in the office of the under secretary of state, May-Sept. 1950, *FRUS, 1950*, 1:580-88.

57. Murray memorandum of conversation with Salazar, 11 Oct. 1951, *FRUS, 1951*, 1:775-76.

58. Murray to Griffis, 19 Oct. 1951, *FRUS, 1951*, 1:778-81.

59. Griffis to Acheson, 19 Dec. 1951, *FRUS, 1951*, 1:793- 94.

60. Groves, *Now*, 182-83.

61. Telephone interview with Joseph Volpe, 10 July 1984.

62. Lilienthal memorandum, 23 April 1947; notes on conversation between Attlee and Smuts, *FRUS, 1947,* 1:804-6, 895-96.

63. Minutes of meeting of American members of CPC, 6 July 1948, *FRUS, 1948,* 1:pt. 2, 719-23. See also Lovett to Hickenlooper, 16 June 1948, in the same volume, pp. 711-12.

64. Minutes of meeting of American members of CPC, 28 May 1948, *FRUS, 1948,* 1:pt. 2, 707-10.

65. Ibid.

66. Arneson to Acheson, 6 Aug. 1949, *FRUS, 1949,* 1:509-11.

67. Arneson to Lovett, 2 Nov. 1948, *FRUS, 1948,* 1:pt. 2, 781-84.

68. Report to Truman by special committee of the National Security Council, 2 March 1949, *FRUS, 1949,* 1:443-61.

69. Arneson to Acheson, 6 Aug. 1949, *FRUS, 1949,* 1:509-11.

70. L. Johnson to Acheson, 13 March 1950, *FRUS, 1950,* 1:542-43.

71. Acheson to L. Johnson, 3 April 1950, *FRUS, 1950,* 1:546-47.

72. Minutes of meeting of American members of CPC and annexes, 25 April 1950; Wendel memorandum of conversation, 12 July 1950, *FRUS, 1950,* 1:547-58, 566-67.

73. Acheson memorandum, 8 Dec. 1950, *FRUS, 1950,* 1:592-93. Donges was former head of the South African Atomic Energy Board. His selection as minister of interior at a time when racial accommodation was a major issue in South Africa and the United Nations suggests the priority which the South African government placed on the uranium sale.

74. Marten to Arneson, 18 Oct. 1950; Arneson to Marten, 4 Dec. 1950, *FRUS, 1950,* 1:589-90, 591-92.

CHAPTER 8: RAISING THE RATE OF COMPENSATION: THE BELGIAN EXPORT TAX

1. Report to Truman by Special Committee of the NSC, 2 March 1949, *FRUS, 1949,* 1:443-61.

2. As quoted in Arneson to Acheson, 13 Sept. 1949, *FRUS, 1949,* 1:526-28.

3. Ibid.

4. In November 1944 Groves had signed a "no profit—no loss" contract with Eldorado for 310 tons of oxide. Though the price paid was above those cited in the Belgian contracts of the same year, the Canadians lost so heavily that in 1946 Groves agreed to pay an additional $3.8 million and to raise the price to $6.17 (Canadian) per pound, in part simply to keep Eldorado functioning. By May of 1947, the price on new Canadian contracts was $13.50 per pound of oxide, yet the Canadians were receiving a profit of only about 10 percent. Comparison of price figures could therefore be quite confusing. Bothwell, *Eldorado,* pp. 189-90, 202-03.

5. Murphy to Acheson, 10 Jan. 1950, *FRUS 1950* 1:493-98.

6. Belgian embassy to department of state, 29 Sept. 1949, *FRUS, 1949,* 1:545-47.

7. Webb to Silvercruys, 5 Oct. 1949, annexed to Arneson to Webb, 5 Oct. 1949, *FRUS, 1949*, 1:554-57.

8. Arneson to Webb, 5 Oct. 1949, *FRUS, 1949*, 1:554-55.

9. Murphy in his dispatch to Acheson of 10 Jan. 1950 quotes from a Sengier letter of 5 Jan. 1950, FRUS, 1950, 1:493-98.

10. The Belgian agenda of 2 Dec. 1949 is quoted in a footnote, ibid.

11. Murphy to Acheson, 10 Jan. 1950, *FRUS, 1950*, 1:493-98.

12. Ibid.

13. Ibid. Telephone interview with Joseph Volpe, 10 July 1984.

14. Murphy to Acheson, 10 Jan. 1950, *FRUS, 1950*, 1:493- 98.

15. Ibid. Also Van Zeeland to Acheson, 17 Feb. 1950, *FRUS, 1950*, 1:525-31.

16. Memorandum on Belgian talks, 18 April 1950, annexed to minutes of meeting of American members of CPC, 25 April 1950, *FRUS, 1950*, 1:547-57.

17. Ibid.

18. Van Zeeland to Acheson, 17 Feb. 1950, *FRUS, 1950*, 1:528-31.

19. Ibid.

20. Ibid.

21. Acheson to Murphy, 22 Feb. 1950, *FRUS, 1950*, 1:531-38.

22. Ibid.

23. Ibid.

24. Ibid.

25. Memorandum on Belgian talks, 28 April 1950, annexed to minutes of meeting of American members of CPC, 25 April 1950, *FRUS, 1950*, 1:547-57.

26. Memorandum of oral communication of Silvercruys, 21 Sept. 1950, *FRUS, 1950*, 1577-78.

27. Ibid. Also letter of Dean F. Frasché to author, 25 May 1983.

28. Memorandum of conversation, 28 Sept. 1950, *FRUS, 1950*, 1:578-80.

29. Arneson to Acheson, 14 Dec. 1950, *FRUS, 1950*, 1:593-96.

30. As quoted in Murphy to Acheson, 17 Nov. 1950, *FRUS, 1950*, 1:590.

31. Arneson to Acheson, 14 Dec.1950, *FRUS, 1950*, 1:593-96.

32. Memorandum of conversation, 12 March 1951, *FRUS, 1951*, 1699-702.

33. Statement of the position of the U.S. and U.K., 5 April 1951, *FRUS, 1951*, 1:715-19.

34. Ibid.

35. Minutes of meeting of U.S., U.K., and Belgian representatives, 5 April 1951, *FRUS, 1951*, 1:713-15.

36. Statement of position of government of Belgium, 1 June 1951, *FRUS, 1951*, 1:730-33.

37. Ibid.

38. Ibid.

39. Minutes of meeting of the U.S., U.K., and Belgian representatives, 11 June 1951, *FRUS, 1951*, 1:733-36.

40. Ibid.

41. Memorandum of conversation, 20 June 1951, *FRUS, 1951,* 1:736-39.

42. Murphy, *Diplomat,* p. 329.

43. Murphy to Acheson, 29 June 1951, *FRUS, 1951,* 1:739.

44. Agreement between the U.S.,U.K., and Belgium, 13 July 1951, *FRUS, 1951,* 1:742-46.

45. Minutes of meeting of secretaries of state, war, and navy, 16 Oct. 1945, *FRUS, 1945,* 2:59-61.

46. Groves to Byrnes, 24 Oct. 1946, *FRUS, 1946,* 1:1258.

47. Gullion to Lovett, 7 April 1948, *FRUS, 1948,* 1:pt. 2, 703-4.

48. Memorandum of conversation, 10 Dec. 1948, *FRUS,* 1948, 1: pt. 2, 791-92. See also Kirk to Marshall, 29 Nov. 1948, *FRUS, 1948,* 3:298-99.

49. Kirk to Lovett, 22 Dec. 1948, Nat. Arch., State, 855A.7962/12-2248.

50. Forrestal to Acheson, 17 Feb. 1949; Kirk to Acheson, 14 March 1949; same to same, 17 March 1949, Nat. Arch., State, 855a.6359/2-1749, /3-1449, /3-1749.

51. Kirk to Acheson, 24 Feb. 1949, Nat. Arch., State, 855A.00/2-2449.

52. FHA to JM, 25 Aug. 1949, Nat. Arch., State, 855a.00B/1-145.

53. Murphy is quoted in a memorandum by Fisher Howe, 9 Sept. 1949, Nat. Arch., State, 855a.00B/8-449.

54. Perkins memorandum of conversation, 25 July 1952, *FRUS, 1952,* 9:pt. 1, 406-9.

55. Murphy telegram of 8 Dec. 1950, as quoted in an editorial note *FRUS, 1950,* 1:590.

56. Murphy to Acheson, 21 Dec. 1949, Nat. Arch., State, 855.002/12-2149.

57. Summary of efforts to improve security of the Belgian Congo, 6 July 1951, *FRUS, 1951,* 1:739-42.

58. Ibid.

59. Memorandum of conversation, 12 March 1951, *FRUS, 1951,* 1:699-702.

60. Summary of efforts to improve security of the Belgian Congo, 6 July 1951, *FRUS, 1951,* 1:739-42.

61. Memorandum of conversation, 25 July 1952, *FRUS, 1952-54,* 11:pt. 1, 406-9.

62. Ibid.

63. Byroade to Cowen, 7 Aug. 1952, *FRUS, 1952-54,* 11:pt. 1, 410-11.

64. Memorandum, 26 Jan. 1953, *FRUS, 1952-54,* 11: pt. 1, 411-12.

65. Telephone interview with Margaret Tibbetts, 14 June 1985, and personal conversations in Brussels, 1961-62, and March 1982.

CHAPTER 9: ATOMS FOR PEACE, ATOMS FOR WAR

1. Taylor and Yokell, *Yellowcake,* p. 26.

2. Hewlett and Duncan, *Atomic Shield,* p. 552. See also Corbin Allardice and Edward P. Trapnell, *The Atomic Energy Commission* (New York: Praeger, 1974), p. 73, and Gordon Dean, *Report on the Atom* (New York: Knopf, 1957), p. 325.

3. Hewlett and Duncan, *Atomic Shield*, p. 74; Thomas L. Neff and Henry D. Jacoby, "The International Uranium Market," Massachusetts Institute of Technology, Laboratory Report No. MIT-EL 80-014, Dec. 1980, Table A-14.

4. Allardice and Trapnell, *Atomic Energy Commission*, p. 42.

5. Silvercruys to Van Zeeland, 1 Oct. 1952, Belgian Ministry of Foreign Affairs, Service des Archives, Collection de Presse, United States, file 12.439. Statement of views of joint chiefs of staff enclosed in Lovett to Lay, 6 Feb. 1952; department of state commentary on JCS paper, 11 June 1952, *FRUS, 1952-54*, 2:pt.2, 863-68; 969-73.

6. Eisenhower quotes a letter to a friend in Dwight D. Eisenhower, *Mandate for Change, 1953-1956* (Garden City: Doubleday, 1963), p. 252.

7. Ibid., pp. 252-53. Lewis L. Strauss, *Men and Decisions* (Garden City: Doubleday, 1962), pp. 357-64.

8. Allardice and Trapnell, *Atomic Energy Commission*, p. 201.

9. Dwight D. Eisenhower, *The Eisenhower Diaries*, edited by Robert H. Ferrel (New York: Norton, 1981), entry for 10 Dec. 1953, pp. 261-62.

10. Ibid., p. 262.

11. Restricted tripartite meeting, 4 Dec. 1953, *FRUS, 1952-54*, 5:pt. 2, 1750-54.

12. Strauss, *Men and Decisions*, p. 357.

13. Arneson to Dulles, 10 March 1953; memorandum of meeting of NSC, 11 March 1953; Strauss to Cutler, 11 June 1953; NSC report, enclosed in Lay note, 13 Aug. 1954, *FRUS, 1952-54*, 2:pt. 2, 1125-33, 1461-62, 1488-99.

14. Reginald Lamarche and Aimé Vaes, "Nuclear Energy in Belgium" *Eurospectra* (June 1966), 5:36-43. Dean, *Report*, pp. 251-83.

15. Dulles to Allen, 24 July 1953, *FRUS, 1952-54*, 11:pt. 2, 1697.

16. Two messages from Allen to Dulles, 26 July 1953, *FRUS, 1952-54*, 11:pt. 2, 1697-99, 1699-1700.

17. Byroade and Waugh to Smith, 31 July 1953, *FRUS, 1952-54*, 11: pt. 2, 1703-05.

18. Allen to Dulles, 1 Aug. 1953, *FRUS, 1952-54*, 11:pt. 2, 1700-1702.

19. Dulles to Allen, 3 Sept. 1953; Allen to Dulles, 8 Sept. 1953, *FRUS, 1952-54*, 11:pt. 2, 1717-19.

20. Smith to Allen, 15 Sept.1953, *FRUS, 1952-54*, 11:pt. 2, 1720-22.

21. Dulles to Allen, 16 Jan.1954, *FRUS, 1952-54*, 11:pt. 2, 1732-35.

22. Allen to Dulles, 29 July 1954, *FRUS, 1952-54*, 11:pt. 2, 1765-66.

23. Smith to Eisenhower, 18 Aug. 1954, *FRUS, 1952-54*, 11:pt. 2, 1712.

24. Jernigan to Smith, 29 Sept. 1954, *FRUS, 1952-54*, 11:pt. 2, 1767-70.

25. Ibid. See also Bishop to Jernigan, 16 Dec. 1954, and fn., *FRUS, 1952-54*, 11:pt. 2, 1796-97.

26. Smith to Johnson, 17 Sept. 1953, *FRUS, 1952-54*, 4:628-31.

27. Ibid.

28. Hamilton memorandum, 13 April 1953; see also Dean to Dulles, 16 Feb. 1953, *FRUS, 1952-54*, 2:pt. 2, 1141-44, 1098-1103.

29. Ibid. and minutes of meeting of U.S. members of CPC, 9 Oct. 1952, *FRUS, 1952-54*, 2:pt. 2, 1026-32.

30. Bothwell, *Eldorado*, p. 254.

31. Ibid., p. 317 and minutes of meeting of American members of CPC, 9 Oct. 1952, *FRUS, 1952-54,* 2:pt. 2, 1026-32.

32. Minutes of meeting of U.S. members of CPC, 9 Oct. 1952; Hamilton memorandum, 13 April 1953, *FRUS, 1952-54,* 2:pt. 2, 1026-32, 1141-44.

33. Johnson report, 4 Jan. 1952; Lee and Thoreson memorandum, 16 Sept. 1952, *FRUS, 1952-54,* 4:902-5, 928-34.

34. Kilcoin to Brown, 17 April 1952, *FRUS, 1952-54,* 4:909-10.

35. Perkins to Acheson, 11 June 1952, *FRUS, 1952-54,* 11:pt. 2, 918-20.

36. Dean to Dulles, 16 Feb. 1953; see also minutes of meeting of U.S. members of CPC, 9 Oct. 1952; Hamilton memorandum, 13 April 1953, *FRUS, 1952-54,* 2:pt. 2, 1098-1103, 1026-32, 1141-44.

37. Strauss to Cutler, 11 June 1954; memorandum of discussion at NSC meeting, 12 Aug. 1954; NSC report, enclosed in Lay note, 13 Aug. 1954, *FRUS, 1952-54,* 2:pt. 2, 1460-61, 1482-99.

38. NSC report, enclosed in Lay note, 13 Aug. 1954, *FRUS, 1952-54,* 2:pt. 2, 1488-99.

39. Minutes of meeting, 12 Nov. 1954, *FRUS, 1952-54,* 2:pt. 2, 1555-58.

40. Smith note of 12 Nov. 1954, *FRUS, 1952-54,* 2:pt. 2, editors' notes, 1558-59.

41. Henry R. Nau, *National Politics and International Technology: Nuclear Reactor Development in Western Europe* (Baltimore: Johns Hopkins Univ. Press, 1974), p. 79fn.

42. Neff and Jacoby, "International Uranium Market," Table A-14. It is difficult from the figures available to determine just how much uranium oxide the Combined Development Trust received from Belgium. The 1200 short tons of 65 percent ore Sengier had stored on Staten Island presumably yielded about 156,000 pounds of oxide. This went directly to the Americans rather than to the CDT. The Americans also held direct option on another 1,000 tons of ore stored above ground in the Congo.

The 1944 Tripartite Agreement provided for delivery of 3,440,000 pounds of oxide to the CDT after which the ten-year term of the agreement would begin. That agreement expired 6 February 1956. It might be assumed, then, that the 3,440,000 pounds of oxides had been delivered by the beginning of February 1946. Yet the Combined Development Trust reported on 26 July 1946 that it had received only 1,143 short tons (2,286,000 lbs.) of oxide prior to 31 March 1946, almost all of which came from the Congo. From 1 April to 26 July 1946 another 1,969 short tons of oxide were received by the CDT. During the last half of 1947 some 1,440 short tons (2,880,000 lbs.) of oxide were obtained by the United States from overseas sources excluding Canada. The Americans at that time were receiving 90 percent of the ores, the British 10 percent. The British therefore gained 160 short tons in the same six months. According to the CDT, all significant shipments of these materials came from the Congo.

No figures are available on production for the last five months of 1946 and for the first six months of 1947. It probably was high in 1946 and began to decline in 1947. Groves had initially estimated an average production rate of 300 short tons per month for the last three quarters of 1946, but by July a near 500 short ton-per-month rate had been achieved. This fell to about 267 short tons per month by the last half of 1947 and would decline to an average of 230 tons per month in 1948. If one conservatively estimates that production for the last five months of 1946 averaged 400 short tons per month and for the first six months of 1947 averaged 300 short tons per month, then the figure deduced for this eleven-month period is 3,800 short tons of oxides. A broad estimate of Belgian uranium oxide received by the Combined Development Trust through the end of 1947 therefore might be 8,512 short tons (1,143 tons prior to 31 March 1946; 1,969 tons April through July 1946; 3,800 tons to 30 June 1947; 1,600 tons July through December 1947). Belgian production from January 1948 through December 1952 was 7,410 metric tons, ranging from 1,300 in 1948 to a high of 2,040 in 1950 and a low of 970 in 1953. This is equivalent to 9,121 short tons of oxides. It therefore appears that the CDT acquired between 17,000 and 18,000 short tons of uranium oxide from Belgian Congo ores in the 1944 to 1953 period. Another 9,759 short tons were produced by the Congo through 1961, after which production was insignificant. These calculations are based on figures obtained from Frank G. Dawson, *Nuclear Power: Development and Management of a Technology* (Seattle: Univ. of Washington Press, 1967), p. 163; Groves to Acheson, 29 April 1946, and report of Groves, Makins, and Bateman, 26 July 1946, *FRUS, 1946*, 1:1238-40, 1254-55; Hewlett and Duncan, *Atomic Shield*, p. 674; Neff and Jacoby, "The International Uranium Market," Table A-14.

CHAPTER 10: PREEMPTION AND MONOPOLY IN RETROSPECT

1. As quoted by Szilard, "Reminiscences," p. 126.

2. Gregg Herken, *The Winning Weapon: The Atomic Bomb in the Cold War 1945-1950* (New York: Knopf, 1980), p. 358, fn. 29.

3. The Franck Report of 11 June 1945 is printed as Appendix B in A. K. Smith, *Peril and Hope*, pp. 560-72.

4. Conant to Bush and Stimson, 30 Sept. 1944, MMRD, H-B, 69.

5. Dean, *Report*, p. 286.

6. Arnold Kramish, *Atomic Energy in the Soviet Union* (Stanford: Stanford Univ. Press, 1959), p. 171.

7. James F. Byrnes, *All in One Lifetime* (New York: Harper, 1958), p. 284.

8. Herken, *Winning Weapon*, pp. 99-113; Martin A. Sherwin, *A World Destroyed: The Atomic Bomb and the Grand Alliance* (New York: Knopf, 1975), pp. 237-38.

9. Gullion to Marshall, 7 Feb. 1948, *FRUS, 1948*, 1:pt. 2, 699-700.

10. Souers to Acheson, Forrestal, and Lilienthal (Pike), 10 Oct. 1949, *FRUS, 1949*, 1:559-64.

11. Groves to Bundy, 6 Feb. 1945, MMRD, H-B, 103.

12. Lovett statement before Joint Congressional Committee, 21 Jan. 1948, *FRUS, 1948*, 1:pt. 2, 688-91.

13. MacArthur memorandum, 2 Dec. 1953, *FRUS, 1952-54*, 5:pt. 2, 1725-26.

14. Irving, *German Atomic Bomb*, pp. 82-83; Arthur H. Compton *Atomic Quest: A Personal Narrative* (New York: Oxford Univ. Press, 1956), p. 97.

15. David E. Lilienthal, *Atomic Energy: A New Start* (New York: Harper and Row, 1980), p. 1. John Simpson, *The Independent Nuclear State: The United States, Britain and the Military Atom* (New York: St. Martin's, 1983), pp. 49-53. David A. Rosenberg, "U.S. Nuclear Stockpile, 1945 to 1950," *The Bulletin of Atomic Scientists* 38 (May 1982), pp. 25-30, and "The Origins of Overkill: Nuclear Weapons and American Strategy, 1945-1960," *International Security*, 7 (Spring 1983):23.

16. Hewlett and Duncan, *Atomic Shield*, p. 415.

17. Simpson, *Independent Nuclear State*, p. 277, fn. 47.

18. Lovett to Lay, 16 May 1952, *FRUS, 1952-54*, 2:pt. 2, 934-37.

19. Minutes of meeting of U.S. members of CPC, 16 April 1952, *FRUS, 1952-54*, 2:pt. 2, 888-94.

20. Statement of special committee of NSC, enclosed in Lay to Lovett and Dean, 10 Sept. 1952, *FRUS, 1952-54*, 2:pt. 2, 1010-13.

21. Minutes of meeting of American members of CPC, 6 July 1948, *FRUS, 1948*, 1:pt. 2, 719-23.

22. Simpson, *Independent Nuclear State*, pp. 1-2.

23. See, for example, the comments of the American joint chiefs of staff, Souers to Acheson, Forrestal, and Lilienthal (Pike), 10 Oct. 1949, *FRUS, 1949*, 1:539-64.

The literature regarding the making of the bomb, the decision for its use in World War II, and its impact on international relations since that war is immense. Yet the amount of information on the search for rare materials for atomic energy purposes is still limited. Many of the published memoirs of key figures of the 1940s and early 1950s contain little or no mention of the scramble for uranium and thorium. Archives stay closed and documents remain classified.

The following is therefore not intended as an encompassing bibliography concerning atomic issues in the decade following the war; nor is it a complete listing of the works consulted in the preparation of this account. It is hoped, however, that it may serve as a brief guide to the diplomacy of the gathering of rare ores.

Primary Sources

DOCUMENTS

Belgium, Ministry of Foreign Affairs, Service des Archives. Collection de Presse. United States.

Brown, Anthony Cave, and MacDonald, Charles, eds. *The Secret History of the Atomic Bomb*. New York: Dial Press/James Waid, 1977.

United States Atomic Energy Commission. *In the Matter of J. Robert Oppenheimer*. Washington: U.S. Government Printing Office, 1954.

United States Congress, House of Representatives Committee on Un-American Activities, 81st cong., *Hearings Regarding Shipment of Atomic Material to the Soviet Union during World War II*. Washington: U.S. Government Printing Office, 1950.

United States Department of State. *Collaboration in Atomic Energy Research and Development*, Publication 5561 (2 pages). Washington: U.S. Government Printing Office, 1954.

United States Office of Strategic Services. Survey of the Belgian Congo, Sept. 1942. U.S. Military History Institute Archives (Carlisle, PA). William Donovan papers.

United States, *Foreign Relations of the United States*. Washington: U.S. Government Printing Office.
United States, National Archives, Department of State files.
  Belgium
  Brazil
  Great Britain
  India
  Portugal
  South Africa
United States, National Archives, Modern Military Records Division.
  Harrison-Bundy papers
  Manhattan Engineering District Decimal Files
  Manhattan Engineering District Top Secret Files
United States Statutes at Large. 79th and 82nd Congress. Washington: U. S. Government Printing Office, 1947, 1951.

DIARIES, MEMOIRS, AND PERSONAL ACCOUNTS

Acheson, Dean. *Present at the Creation: My Years in the State Department*. New York: Norton, 1969.
Byrnes, James F. *All in One Lifetime*. New York: Harper, 1958.
Compton, Arthur H. *Atomic Quest: A Personal Narrative*. New York: Oxford Univ. Press, 1956.
Dean, Gordon E. *The Reminiscences of Gordon Dean*. Columbia University Oral History Research Office, 1959. New York: Microfilming Corporation of America, Glen Rock, N. J., 1972.
———. *Report on the Atom*. 2nd ed. New York: Knopf, 1957.
Eisenhower, Dwight D. *The Eisenhower Diaries*. Edited by Robert H. Ferrell. New York: Norton, 1981.
———. *Mandate for Change, 1953-1956*. New York: Doubleday, 1963.
Groves, Leslie R. *Now It Can Be Told*. New York: Harper, 1962.
Gutt, Camille. *La Belgique au Carrefour, 1940-1944*. Paris: Fayard, 1971.
Kennan, George F. *Memoirs 1925-1950*. Boston: Little Brown, 1967-72.
Lilienthal, David E. *Atomic Energy: A New Start*. New York: Harper and Row, 1980.
———. *Change, Hope, and the Bomb*. Princeton: Princeton Univ. Press, 1963.

———. *The Journals of David E. Lilienthal.* Vol 2: *The Atomic Energy Years, 1945-1950.* New York: Harper and Row, 1964.

Murphy, Robert. *Diplomat Among Warriors.* New York: Doubleday, 1964.

Strauss, Lewis L. *Men and Decisions.* Garden City: Doubleday, 1962.

Spaak, Paul-Henri. *Combats inachevés.* 2 vols. Paris: Fayard, 1969.

Szilard, Leo. "Reminiscences," edited by Gertrude W. Szilard and Kathleen R. Winsor. In *The Intellectual Migration: Europe and America, 1930-1960,* pp. 94-151. Edited by Donald Fleming and Bernard Bailyn. Cambridge, Mass.: Harvard Univ. Press, 1969.

Secondary Sources

Allardice, Corbin, and Trapnell, Edward P. *The Atomic Energy Commission.* New York: Praeger, 1974.

Bothwell, Robert. *Eldorado: Canada's National Uranium Company.* Toronto: Univ. of Toronto Press, 1984.

Cervenka, Zdenek, and Rogers, Barbara. *The Nuclear Axis: Secret Collaboration between West Germany and South Africa.* New York: Times Books, 1978.

Clark, Ronald W. *The Greatest Power on Earth: The International Race for Nuclear Supremacy.* New York: Harper and Row, 1980.

Dawson, Frank G. *Nuclear Power: Development and Management of a Technology.* Seattle: Univ. of Washington Press, 1976.

Del Sesto, Steven L. *Science, Politics, and Controversy: Civilian Nuclear Power in the United States, 1946-1974.* Boulder: Westview Press, 1979.

Durie, Sheila, and Edwards, Rob. *Fuelling the Nuclear Arms Race.* London: Pluto Press, 1982.

Eggleston, Wilfrid. *Canada's Nuclear Story.* Toronto: Clarke, Irwin, 1965.

Ford, Daniel. *The Cult of the Atom: The Secret Papers of the Atomic Energy Commission.* New York: Simon and Schuster, 1982.

Fox, Annette Baker. *The Power of Small States: Diplomacy in World War II.* Chicago: Univ. of Chicago Press, 1959.

Gowing, Margaret. *Britain and Atomic Energy 1939-1945.* United Kingdom Atomic Energy Authority. New York: St. Martin's, 1964.

Gowing, Margaret, assisted by Arnold, Lorna. *Independence and De-*

terrence: *Britain and Atomic Energy, 1945-1952.* Vol. 1: *Policy Making.* Vol. 2: *Policy Execution.* New York: St. Martin's, 1974.

Grishin, Nikolai. "The Saxony Uranium Mining Operation ('Vismut')." In *Soviet Economic Policy in Postwar Germany: A Collection of Papers by Former Soviet Officials,* pp. 127-53. New York: Research Program on the U.S.S.R., 1953.

Harris, Kenneth. *Attlee.* New York: Norton, 1982.

Helmreich, Jonathan E. *Belgium and Europe: A Study in Small Power Diplomacy.* The Hague: Mouton, 1976.

———. "The Uranium Negotiations of 1944." In *Le Congo Belge durant la Seconde Guerre Mondiale.* Brussels: Académie royale des sciences d'outre-mer/Koninklijke Academie voor Overzeese Wetenschappen, 1983, pp. 253-83.

Herkin, Gregg. *The Winning Weapon: The Atomic Bomb in the Cold War 1945-1950.* New York: Knopf, 1980.

Hewlett, Richard G., and Anderson, Oscar E. *The New World, 1939/ 1946.* Vol. 1 of *A History of the United States Atomic Energy Commission.* 2 vols. University Park, Pa.: Pennsylvania State Univ. Press, 1962-69.

Hewlett, Richard G., and Duncan, Francis. *Atomic Shield, 1948/ 1952.* Vol. 2 of *A History of the United States Atomic Energy Commission.* 2 vols. University Park, Pa.: Pennsylvania State Univ. Press, 1962-69.

Irving, David. *The German Atomic Bomb: The History of Nuclear Research in Nazi Germany.* New York: Simon and Schuster, 1967.

Knebel, Fletcher, and Bailey, Charles W. II. *No High Ground.* New York: Harper, 1960.

Kohl, Wilfrid L. *French Nuclear Diplomacy.* Princeton: Princeton Univ. Press, 1971.

Kramish, Arnold. *Atomic Energy in the Soviet Union.* Stanford: Stanford Univ. Press, 1959.

Lamarche, Reginald, and Vaes, Aimé. "Nuclear Energy in Belgium." *Eurospectra* 5 (June 1966):36-43.

Nau, Henry R. *National Politics and International Technology.* Baltimore: Johns Hopkins Univ. Press, 1974.

Neff, Thomas L., and Jacoby, Henry D. "The International Uranium Market." Massachusetts Institute of Technology Energy Laboratory Report No. MIT-EL-80-014. December 1980.

Nieburg, Harold L. *Nuclear Secrecy and Foreign Policy.* Washington: Public Affairs Press, 1964.

Pash, Boris T. *The ALSOS Mission*. New York: Award Books, 1969.

Pringle, Peter, and Spigelman, James. *The Nuclear Barons*. New York: Holt, Rinehart, and Winston, 1981.

Rogow, Arnold A. *James Forrestal: A Study of Personality, Politics, and Policy*. New York: MacMillan, 1963.

Rosenberg, David A. "The Origins of Overkill: Nuclear Weapons and American Strategy, 1945-1960." *International Security* 7 (Spring 1983):3-71.

———. "U. S. Nuclear Stockpile, 1945 to 1950." *The Bulletin of the Atomic Scientists* 38 (May 1982):25-30.

Scheinman, Lawrence. *Atomic Energy Policy in France under the Fourth Republic*. Princeton: Princeton Univ. Press, 1965.

Sherwin, Martin J. *A World Destroyed: The Atomic Bomb and the Grand Alliance*. New York: Knopf, 1975.

Simpson, John. *The Independent Nuclear State: The United States, Britain and the Military Atom*. New York: St. Martin's, 1983.

Smith, Alice Kimball. *A Peril and a Hope: The Scientists' Movement in America: 1945-57*. Chicago: Univ. of Chicago Press, 1965.

Wheeler-Bennett, John W. *John Anderson, Viscount Waverley*. New York: St. Martin's, 1962.

INDEX

Acheson, Dean, 160, 171, 178, 187–88, 193, 200–1, 206, 215, 220, 243, 251, 261–2; and Belgian wish to publicize the Tripartite Accord, 80, 82, 85–90; discussion with British on information exchange, 110–12, 115–17, 120–22, 130, 139–40, 143–46, 148–53, 155–56

Achilles, Theodore C., 87

Advisory Committee on Uranium (United States), 5, 8, 9

African Metals Corporation, 6, 18, 29, 32, 33, 38, 73, 77, 83

Agreement and Declaration of Trust, 15

aide-mémoire of 1950, 206, 208

Akers, W. A., 99–101

Alexander, A. V., 136

Algoma, 225

Allardice, Corbin, 228–29

Allen, George V., 232, 234

Anderson, Sir John, 11–13, 53, 68, 73, 178, 250, 258, 261–62; and information-sharing controversy, 99, 107–8, 119; negotiations with Belgium in 1944, 15–18, 20–27, 30–31, 38–39, 265n; negotiations with the Netherlands, 57–59; negotiations with Sweden, 61–62, 64–65

Anderson, Clayton and Company, 54

Anseele, E., 78

apartheid, 242

Argentina, 69, 166, 239

Armed Forces Special Weapons Command (United States), 122

Arneson, R. Gordon, 106, 137, 139, 143, 153, 164, 176, 179, 187, 223–24, 251; and Belgian export tax of 1951, 194–95, 197, 199, 206, 208, 214

Athabasca, 240

atomic bomb, xiii, 10, 41–42, 70–71, 88, 106, 108–9, 125, 127, 154–55, 248, 253, 255–56; British and Soviet production and testing of, xiii, 5, 7, 9, 12, 99, 102, 135–36, 140–42, 144–45, 149, 152–53, 157, 173–74, 179, 183, 195, 197, 207, 230, 249, 256, 259, 261

atomic energy: commercial and industrial use, xvii, 53, 55, 58, 70–72, 79, 81, 86–89, 91, 95, 100, 106, 111, 164–65, 167, 172, 175, 177, 192, 194–98, 202–3, 208, 212, 229, 244, 258, 262; military-civilian debate over control, xi, 105, 109, 119–20, 122, 134

Atomic Energy Act of 1946 (McMahon Act), 85–86, 88, 95, 110, 124–25, 128–30, 137, 139–40, 144, 153–54, 178, 182, 199, 201–2, 231, 243, 251–52; amendment in 1951, 145–52; passage, 81–82, 112, 118–21; revision in 1954, 226–31, 244

Atomic Energy Act (India), 169

Atomic Energy Board (South Africa), 188–89, 280n

Atomic Energy Commission (Belgium), 80, 207

Atomic Energy Commission (France), 171, 179–83

Atomic Energy Commission (India), 169–70, 234

Atomic Energy Commission (United States), 86–88, 90, 92, 95, 103, 105, 109–10, 112, 158, 160, 164–65, 168, 171, 175, 188, 193, 195–96, 199, 213–14, 225–26, 229, 230, 234, 236–41, 243–47, 261; creation, 85; information-sharing and ore-allocation controversies, 119, 121–22, 124, 134, 137, 139, 143–44, 146, 151–52

Index

Eisenhower, Dwight D., 19, 111, 120–22, 132, 139–40, 157, 193, 244, 261–62; and Atoms for Peace Program, 227–29, 243, 246, 250

Eldorado Gold Mines, Ltd., 6, 8, 11–12, 18, 45–47, 60, 99, 104, 115, 197, 240–41, 263n, 280n

electricity, 18, 35, 91–92, 222, 226, 242, 258, 265n

Elisabethville, 218–19

Emergency Procurement Service (United States), 237

Empresa Mineira concession, 68

Eniwetok, 136, 145, 148–49, 154

espionage, xii, 88, 96, 143, 148, 152, 255. *See also* Fuchs; Gouzenko spy ring; Maclean; May

Espirito Santo, 51

Estonia, 255

European Nuclear Research Laboratory, 180, 182, 230

European Recovery Program, 89–91, 125, 127–28, 131–32, 136, 139, 141, 157, 162, 174, 182, 261

Eyskens, Gaston, 193–94, 202, 204

Federal Reserve Bank, 9, 162

Feis, Herbert, 7

Fermi, Enrico, 4, 5

Fernandes, Raul, 161

Finletter, Thomas K., 7

Fisher, Adrian, 143

Fountoura, Neves da, 164–65

Force Publique, 219, 221

Foreign Operations Administration (United States), 236

Forrestal, James V., 104, 109, 135–36, 139, 178, 187, 193, 218

Foster, William C., 152

France, xi, 30, 37, 97, 102, 158–59, 161, 164, 170–72, 176–82, 184, 190, 200, 210, 230, 241, 245, 250, 257

Franck, James, 248

Franco, Francisco, 184–85, 190

Franks, Oliver, 155–56, 215

Free University of Brussels, 76, 84, 94

Freeman, William, 237

French Equatorial Africa, 181

French Morocco, 178, 181–82, 241

French Union, 181

Fuchs, Klaus E. J., 101, 143–45, 183, 188, 192, 256

gas diffusion, 43, 111, 230, 264n

Gaudin, A. M., 185

Geheniau (Belgian professor), 84

Geiger-Muller counter, 44, 225

General Electric Corporation, 164–65

Gerard, Paul, 94

German Democratic Republic, 70

Germany (Third Reich), 4, 5, 7, 25, 37, 42, 45, 49, 51, 58–59, 113, 191, 247, 249, 252, 254

Godding, Maurice, 75, 78, 80, 85, 89

Gold, Harry, 143

gold mining, 6, 70, 185

Gomes, Eduardo, 159

Gorell Barnes, W. L., 16–17, 19, 27

Gouzenko spy ring, 112, 118–19, 130

Gowing, Margaret, xii, 107, 130–31

Grants, N. M., 225

Great Bear Lake, 8, 43

Great Britain, *see* United Kingdom

Greece, 84, 119–20, 123

Greenglass, David, 143

Groves, Leslie R., xi, 5–6, 8–10, 13–15, 22, 27, 29, 53, 69–71, 128, 134, 144, 159, 217, 223, 237, 247–53, 258–59, 262, 264n, 280n, 284n; and African Metals Corporation contract of September 1944, 33–35, 37, 40, 42–43, 46, 48–49; differences with the British over information exchange, 98–99, 102–18; negotiations with Brazil, 49–57 *passim*; negotiations with the Netherlands, 57–59; negotiations with Sweden, 61–67 *passim*; and price talks with the Belgians, 1945–46, 73–79; retirement, 121–23

Guarin, Paul L., 43

Gueben (Belgian professor), 95

Gullion, Edmund, 82–83, 85, 87, 94, 123–24, 160, 178, 218, 251–52

Gustafson, John K., 170, 199, 240

Gutt, Camille, 17–18, 21, 26–28, 30, 35, 38–39, 41, 96, 257–58

Halban, Hans van, 102–3, 178

Halifax, Lord, 57, 104, 112–14, 116

Library of Congress Cataloging-in-Publication Data

Helmreich, Jonathan E.
  Gathering rare ores.

  Bibliography: p.
  Includes index.
  1. Nuclear energy—Government policy—United States—History.   2.
Nuclear industry—Political aspects—History.   3. Uranium industry—
Political aspects—History.   4. United States—Foreign relations—Great
Britain.   5. Great Britain—Foreign relations—United States.   6.
World politics—1945-   I. Title.
HD9698.U52H45   1986   327.1'11   86-4862
ISBN 0-691-04738-3   (alk. paper)

Jonathan E. Helmreich is Professor of History at
Allegheny College and the author of *Belgium and
Europe: A Study in Small Power Diplomacy* (Mouton).